D0932033

Structural and Depositional Styles of Gulf Coast Tertiary Continental Margins: Application to Hydrocarbon Exploration

Continuing Education Course Note Series # 25

**Martin P.A. Jackson and
William E. Galloway**
Bureau of Economic Geology,
The University of Texas at Austin

This book is an author-prepared publication
of the AAPG Education Program.
Extra copies of this, and other titles in the
Education Course Note Series, are available from:

The AAPG Bookstore
P.O. Box 979
Tulsa, OK 74101, USA

Published April 1984
ISBN: 0-89181-174-5

COURSE OUTLINE

PRINCIPLES

APPLICATION TO GULF COAST TERTIARY

INTRODUCTION AND ACKNOWLEDGMENTS

This course is the outgrowth of our increasing recognition of the fact that the structure and genetic stratigraphy of the Gulf of Mexico continental margin are, like the chicken and the egg, inextricably intertwined. As hydrocarbon exploration and exploitation advance to and beyond the present shelf edge and into the deeply buried Tertiary basin fill, interpretation of the complex depositional and structural styles of the outer shelf and upper slope setting will increasingly challenge the interpreter. Our hope is that these notes will provide a coherent summary of the key concepts, models, and tools that are needed to meet this exploration challenge. Material is presented in the form of a series of short chapters or units, each dealing with a specific topic and containing numerous illustrations and citations of key references for further reading. Though the units are written so as to be self-contained, they are arranged in what we believe is a logical sequence leading to a general synthesis of exploration concepts.

We gratefully acknowledge the support of the Bureau of Economic Geology of The University of Texas at Austin in the preparation of these notes. Early versions were typed by Marcia J. Franklin and Virginia C. Zeikus. Amanda R. Masterson edited and managed the compilation of the final manuscript. Word processing was done by Joann Haddock and Dorothy C. Johnson and figures were drafted under the supervision of Richard L. Dillon. Margaret L. Evans and Jamie A. McClelland pasted up the final manuscript. To all who contributed to the preparation of these notes, we express our appreciation for a job so typically well done.

CONTINENTAL MARGINS: BASIC PRINCIPLES

The continental margin, as the term is commonly used, encompasses a critical boundary between the shallow to subaerial environments that rim a basin and the deep basin floor. Also implicit in the term is the association with a crustal boundary or transition separating continental and oceanic basement. Thus, continental margins are major physiographic, bathymetric, and structural features of the Earth's skin.

The three bathymetric and depositional regimes that form the continental margin include the shelf, shelf edge, and continental slope (fig. 1). As we shall see, our concepts about continental shelves, which are prejudiced by the Holocene high stand of sea level, may require some modification.

The modern continental margin of the Northwest Shelf of the Gulf of Mexico is readily described in terms of water depth and depositional gradient. However, recognition of paleocontinental margins in the stratigraphic record proves more difficult. Several criteria that may be used include (1) paleobathymetry, (2) seismic stratigraphy, and (3) syndepositional structural growth. The latter two criteria, as recognized in the surficial Quaternary sediments of the Gulf of Mexico (offshore of Galveston), are shown in figure 1.

Key Reference: Winker (1982)

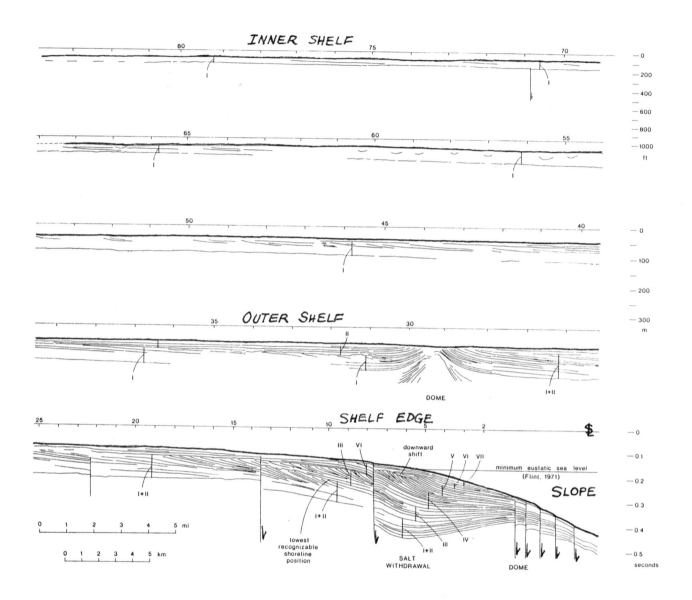

Figure 1. Tracing of a sparker profile from Galveston across the shelf edge. Vertical exaggeration approximately 18x. Important features include (1) gradual basinward thickening of individual seismic sequences (labeled with roman numerals), (2) low-angle clinoforms across the outer shelf, (3) thickening, steepening, and greater separation into distinct depositional sequences of clinoforms at the shelf edge, (4) major growth faults at or near the shelf edge, (5) extreme lateral thickness variations of depositional units, particularly at the upper slope, and (6) penetration of the section by piercement (outer shelf) and deep-seated (upper slope) salt domes. From Winker (1979).

QUATERNARY PERSPECTIVES

Our perception of continental margin bathymetry and stratigraphy is strongly influenced by the fact that we are looking at continental margins that underwent extensive, geologically abrupt, eustatic transgresion that ended only a few thousand years ago. Because of this eustatic rise in sea level, totaling more than 400 ft (120 m), depositional basin margins, such as the northern Gulf of Mexico, now have broad, low-relief continental shelves. The modern shelf edge typically lies (or is picked) at the 600 ft (180 m) bathymetric contour.

However, the active growth of modern continental margins occurred during the extensive periods of low sea level throughout the Quaternary. The continental shelves are largely transgressed coastal plains composed of fluvial, deltaic, shore-zone, and narrow shelf deposits (fig. 1). Stripping away the transgressive Holocene veneer reveals a depositional platform consisting of:

1. Deltaic headlands
2. Narrow prodelta shelves
3. Interdeltaic shore-zones and relatively steep shelves morphologically similar to the carbonate ramp
4. Fronting continental slopes

The focus of deposition, major deltaic headlands, prograded rapidly to the shelf edge. Once at the shelf edge, the deltas prograded directly onto the continental slope. The prodelta and upper slope are thus one and the same environment.

Key References: Lehner (1969); Sidner et al. (1978); Winker (1979)

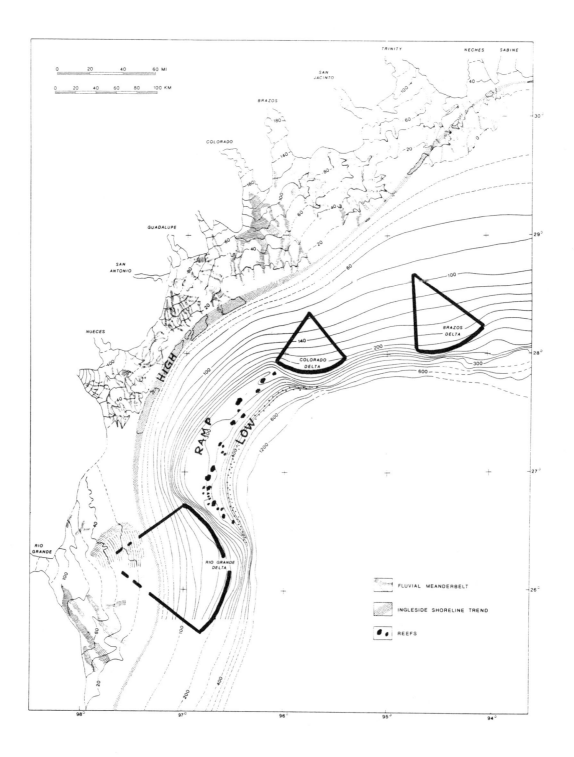

Figure 1. Paleotopography and bathymetry of the Texas continental shelf at the end of Pleistocene time. Contours were obtained by subtracting the thickness of Holocene sediment (principally shelf mud). Approximate high and low stands of Quaternary sea level are indicated by stippled belts just inland of the modern, largely transgressive shoreline and along the present shelf edge. Modified from Winker (1979).

DELTA SYSTEMS I

A delta is a river-fed depositional system that produces an irregular progradation of the shoreline. The entire, subaerial and subaqueous, contiguous sediment mass is included in the delta system. If this basic concept is accepted, many of the arguments regarding depositional environments of continental margin reservoir sandstones are readily resolved.

Channel Mouth Processes

Deposition of sediment at the distributary channel mouth is the fundamental process of delta systems (fig. 1).

1. Deposition of channel mouth bar
2. Generation of frontal splays
3. Deposition of prodelta mud platform
4. Reworking of channel mouth bar by waves and tidal currents

Gravitational Resedimentation and Deformation

Channel-mouth and delta-front sediments are ideally situated for modification, disruption, or remobilization by gravitational potential energy. First, sand is deposited on top of a thick, undercompacted, water-saturated, prodelta mud platform. Second, the locus of deposition lies at the crest of a sloping prodelta and upper slope apron.

The modern lobe of the Mississippi delta is prograding onto thick prodelta muds at the crest of the modern continental slope. The various observed deformational features include (fig. 2):

1. Slumps and larger faults
2. Mud flows (which can trigger turbidity currents)
3. Mud diapirs

The close physical and genetic association of growth faults and slumps with the delta-front sand facies is repeatedly found at a variety of scales, ranging from thin, intracratonic basin deltaic sequences to thick deltas such as the Triassic of Svalbard (fig. 3).

Key References: Coleman and Prior (1980); Galloway and Hobday (1983)

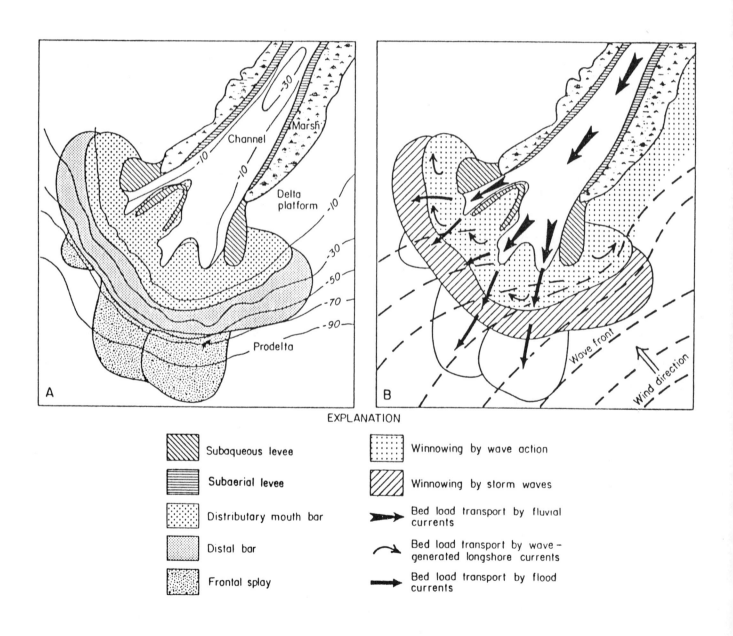

Figure 1. Channel-mouth environments and depositional processes. From Galloway and Hobday (1983).

6

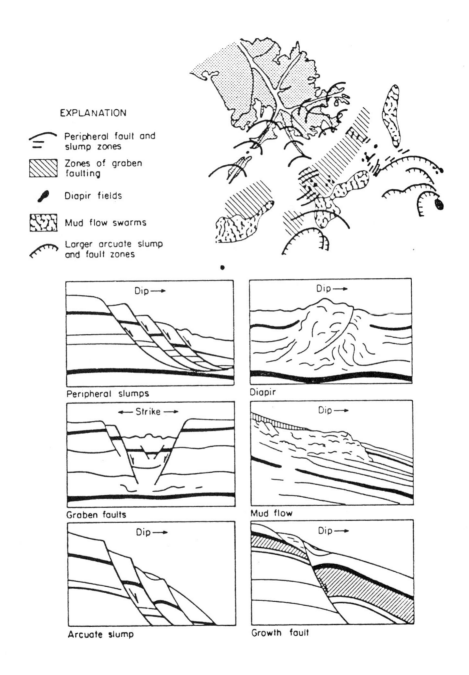

EXPLANATION

◠ Peripheral fault and
‖ slump zones

▨ Zones of graben
faulting

◗ Diapir fields

▧ Mud flow swarms

◠ Larger arcuate slump
and fault zones

Peripheral slumps

Diapir

Graben faults

Mud flow

Arcuate slump

Growth fault

Figure 2. Styles of gravitational resedimentation and deformation along an actively prograding delta front. From Galloway and Hobday (1983); modified from Coleman and Garrison (1977).

Perspective

Single fault block

Figure 3. Triassic delta-front growth faults, Svalbard, Norway. From Galloway and Hobday (1983); courtesy of M. B. Edwards.

DELTA SYSTEMS II

Review of the process framework and associated morphology and sediment distribution patterns in modern deltas shows that three basic process suites determine delta geometry and distribution of framework sand facies: (1) sediment input, (2) wave energy flux, and (3) tidal energy flux. Major remobilization and displacement of bed-load sediment from the delta front may generate a coupled, but depositionally distinctive, submarine slope system. A tripartite classification of delta systems, based on dominance of depositional process, is shown in figure 1.

Sand-Body Geometry and Vertical Sequences

Only fluvial- and wave-dominated delta systems have been shown to be major components of the Gulf Coast Tertiary basin fill.

Framework sand-body geometry of each delta type is quite different (fig. 2). The fluvial-dominated delta produces digitate, bifurcating mouth bars cored by distributary channels (fig. 2A). Orientations are highly variable, but are generally dip-directed. Vertical sequences reveal a highly irregular, upward-coarsening pattern (fig. 3). Upward increase in textural size is largely due to the increasing thickness and abundance of sand beds. Frontal splay "turbidites," slump blocks, and sediment flow units are all abundant, especially in continental margin deltas where deposition was directly onto the crest of the slope.

Wave-dominated deltas show similar vertical sequences near active channel mouths but produce thick, commonly massive coastal barrier sand bodies in interdistributary delta-front positions (fig. 4). The extensive wave reworking produces a laterally continuous, irregularly strike-oriented sand body (fig. 2B) dominated by the coastal barrier facies.

Reservoir Facies

Reservoir facies of fluvial-dominated delta margins consist of the channel mouth bar and distributary channel sand bodies. Largest slump deposits may have sufficient volume to be local reservoirs. Thick sequences (to be discussed later) of distal delta front sands (slump and frontal splay deposits) may provide discontinuous, low-permeability reservoirs in the Gulf Coast Tertiary.

Reservoir facies of the wave-dominated deltas are primarily the laterally continuous coastal barrier sands. Frontal splay and slump facies may also be locally important, generally low-permeability gas producers. Ease of field development increases as importance of wave reworking of the delta front increases.

Key References: Galloway (1975); Galloway and Hobday (1983)

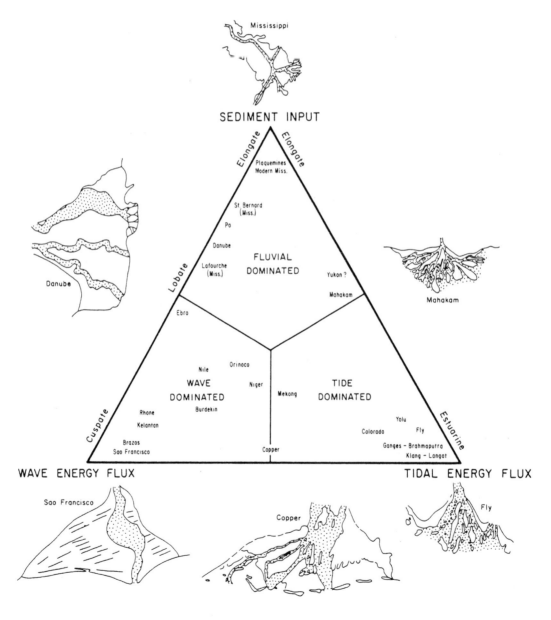

Figure 1. Processes and morphologic classification of delta systems. From Galloway and Hobday (1983).

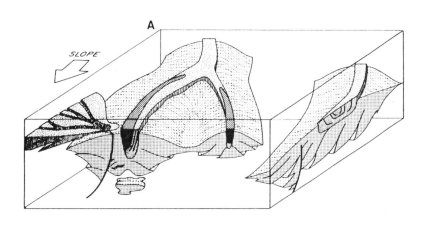

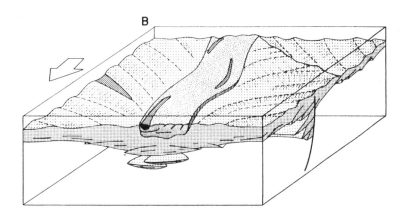

Figure 2. Framework sand architecture of delta systems. A. Fluvial-dominated delta. B. Wave-dominated delta. From Galloway and Hobday (1983).

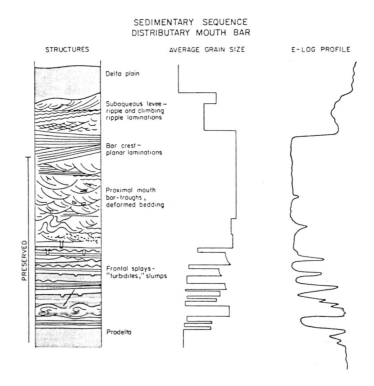

Figure 3. Generalized vertical profiles through a channel-mouth bar sand body. From Galloway and Hobday (1983).

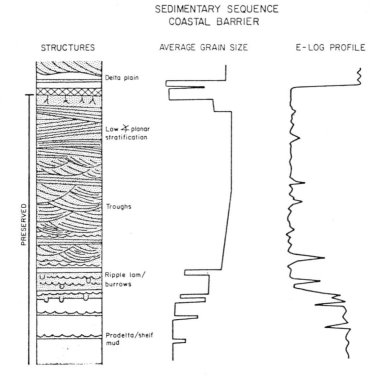

Figure 4. Generalized vertical profiles through a coastal-barrier sand body of a wave-dominated delta margin. From Galloway and Hobday (1983).

SHELF SYSTEMS

Shelf processes most effective in transporting bed-load sediment are related to tides, storm waves and surge, and wind-forced currents.

In a Mediterranean basin, such as the Gulf of Mexico, tide range is limited and <u>tidal</u> <u>processes</u> have likely been of minor importance.

Below fair-weather wave base, which establishes the depth of the shoreface (typically about 30 ft or 10 m in modern and Tertiary Gulf Coast), <u>storm waves</u> may entrain and transport sediment to depths of a few hundred feet. Hummocky crossbedding is now popularized as a key sedimentary structure indicative of storm wave action on the shelf.

<u>Storm surge</u> occurs when wind and barometric pressure pile water against the shore. With passage of the storm, return flow also generates bottom currents capable of sand transport.

<u>Wind-forced currents</u> are produced at shallow to moderate depths as storm winds pile water against the shore. Unlike storm surge, flow may continue for several days. Flow is deflected along strike by the shoreface. Combined offshore transport (ebb surge) and strike transport (wind-forced currents) may play a major role in sand deposition in water depths below normal current action and at the shelf edge. Such processes rapidly lose their impact in deeper water beyond the shelf edge, however.

Depositional Record

In the Gulf Tertiary Basin, both wind-forced currents and storm surge have produced sand units within otherwise mud-rich shelf depositional systems. The resultant system may be described as storm punctuated (fig. 1). A generalized vertical sequence produced by progradation of such a storm-punctuated shelf system is shown in figure 2. The expanded upward-coarsening sequence consists of multiple, thin, amalgamated storm beds that increase in thickness and abundance upward (in shallower water). The sequence merges up into the lower shoreface sands of a delta front or shore zone.

<u>Key References:</u> Galloway and Hobday (1983); Morton (1981)

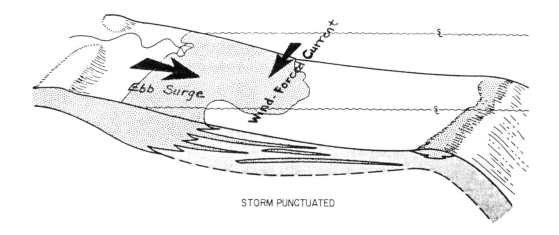

STORM PUNCTUATED

Figure 1. Sand distribution in a storm-punctuated shelf system.

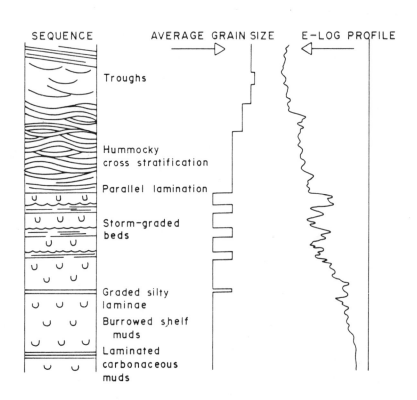

Figure 2. Upward-coarsening sequence produced by progradation of a storm-punctuated shelf. From Galloway and Hobday (1983).

SUBMARINE SLOPE SYSTEMS I

Submarine slopes are characterized by their relatively steep declivity. Although the angle of the slope is commonly less than $5°$ in active depositional slopes, this is sufficient to result in considerable instability. The upper slope in particular is a zone of sediment instability and bypass, and is a focus of erosional processes. Sediment remobilization and transport are dominated by gravity mass-transport and density underflow processes. Mass-transport processes include:

1. Slumping along discrete glide planes
2. Debris flows of viscous sediment-water mixtures

Because gravitational potential energy is the principal driving mechanism for down-slope transport, slope systems are distinguished by their inherent tendency to deposit coarsest sediment at the base of the depositional sequence in bathymetric lows (fig. 1). Sand, though present as a part of the resedimented material, is unlikely to be deposited at the crest of bathymetric highs, as shown by core results from the Quaternary slope of the Northern Gulf (table 1).

Sediment contained in the slope depositional unit reflects the material deposited at the shelf edge and upper slope. For example, slumping of delta front sands (an extremely common event in shelf-edge deltas) may displace large masses of sand down the upper slope (fig. 2). A single large slump may initiate submarine canyons, which erode headward and backfill their lower parts with debris flow and turbidite sediment (fig. 3). Recent work shows that the large canyons and gorges associated with the modern Mississippi have such a slump-dominated origin.

Depositional Models

Two depositional models have been applied to the Gulf Coast Tertiary sedimentary fill. The well-known submarine fan model has been used many times. Elements of this model are summarized in figures 4 and 5. Key components include an upper fan channel and a lower aggradational apron of suprafans. This model was developed from studies of tectonically active, sand-rich, uplap slopes. Though the modern Mississippi fan shares some elements with the classic submarine fan model, this model does <u>not</u> appear to be a particularly useful one for most of the mud-rich, unstable, prograding Gulf of Mexico continental margin.

The Quaternary Gulf slope provides alternative models of slope systems. Much of the slope consists of a complex of discrete depositional layers, lobes, wedges, and aprons of debris flow and slump material. Locally, slumping and erosion combine to produce large submarine canyons or gorges that fill with slump and confined turbidite fan deposits (fig. 3).

Framework sand-body geometries of the sand-rich submarine fan and gorge fill are illustrated in figure 6, A and C. Slumping and debris flow do not produce an integrated framework-sand distribution system, but considerable sand may be deposited on the slope by such processes.

Table 1. Sand Recovery from Various Bathymetric Settings*

<u>Bathymetric setting</u>	No. of <u>core tests</u>	No. with <u>sand</u>	% with <u>sand</u>
Closed high	3	0	0
Closed depression	3	3	100
Flank of high or ridge	7	1	14
Region of slope	16	4	25
	29	8	

*Where recovery was 1.5 m or more.
From Woodbury et al. (1978)

<u>Key</u> <u>References</u>: Coleman and Prior (1983); Galloway and Hobday (1983); Moore et al. (1978); Walker (1978); Watkins and Kraft (1978); Woodbury et al. (1978)

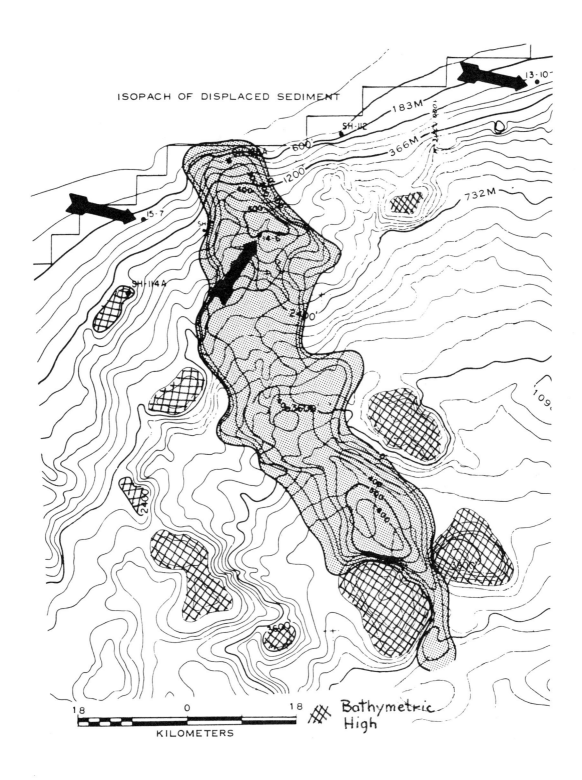

ISOPACH OF DISPLACED SEDIMENT

Bathymetric High

18   0   18

KILOMETERS

Figure 1. Isopach map of slump-displaced sediments superimposed on a bathymetric map. Note emplacement of debris-flow sediments between diapir-produced bathymetric highs. Pleistocene, Texas shelf margin. Modified from Woodbury et al. (1978).

17

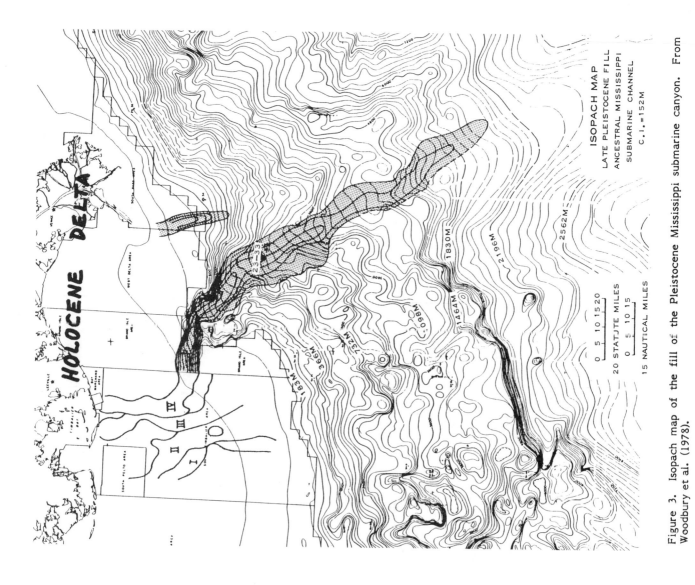

Figure 3. Isopach map of the fill of the Pleistocene Mississippi submarine canyon. From Woodbury et al. (1978).

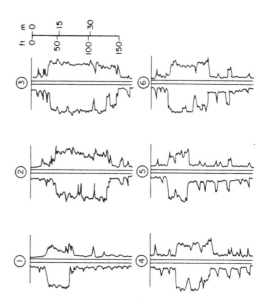

Figure 2. Large slump deposits of the Mississippi delta front. From Galloway and Hobday (1983); modified from Coleman and Pryor (1980).

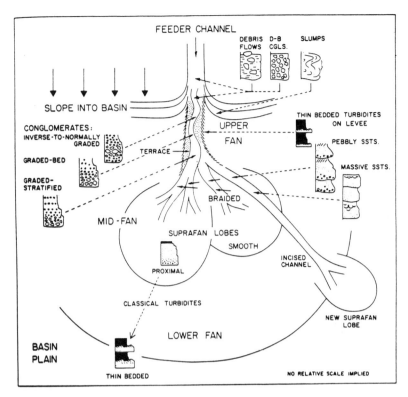

Figure 4. Sand-rich submarine fan model showing fan morphology, depositional environments, and facies distribution. From Walker (1978).

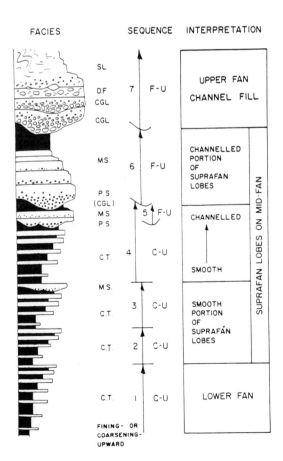

Figure 5. Composite stratigraphic sequence developed by sand-rich fan offlap. C-U = coarsening-upward. F-U = fining-upward. From Walker (1978).

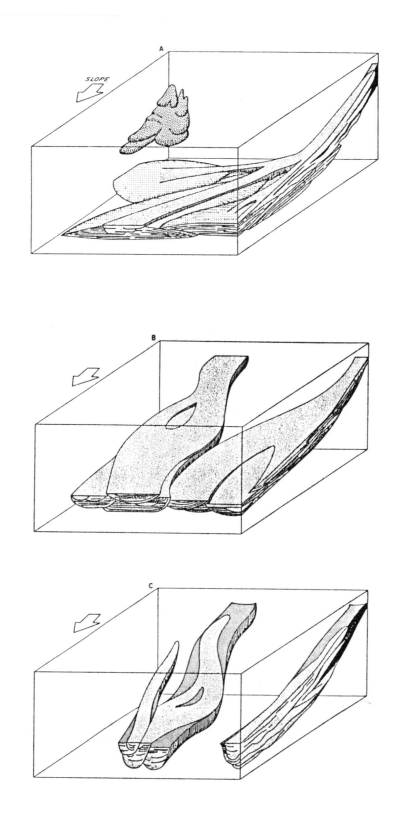

Figure 6. Framework sand architecture of submarine slope systems. A. Sand-rich fan and slump. B. Nonturbid density current fan (not important in Gulf). C. Onlap submarine gorge fill. From Galloway and Hobday (1983).

SUBMARINE SLOPE SYSTEMS II

Seismic stratigraphy has proved particularly useful in delineating the depositional architecture of continental margins. The scale of slope units is commonly adequate for resolution by CDP data, and the depositional topography of slope features -- canyons, fans, slump sheets, etc. -- is diagnostic.

The relationship of slope bedding to bounding stratal surfaces or features is particularly useful for distinguishing major types of slope systems. Basic slope architectures are (fig. 1):

(1) Offlap (progradational)

(2) Onlap

(3) Uplap

Most long-lived depositional continental margins, including the Gulf of Mexico, display an alternation between offlap and onlap depositional styles (fig. 2). Slope offlap results in outbuilding or construction of the continental platform. Onlap, in contrast, commonly reflects sediment bypass and erosion of the continental margin. Sediment reworked from the outer shelf, shelf margin, and upper slope is redeposited along the lower slope, reducing the slope gradient.

The depositional pattern of a simple (structurally stable) basin margin is illustrated by many late Paleozoic basin margins (fig. 3). Sand distribution is bimodal -- deltaic and shore-zone sands cap the offlap units and submarine fan sands are concentrated in the lower slope and adjacent basin floor.

The Gulf Tertiary margin, however, is structurally unstable. Progradation onto salt and/or underlying thick, underconsolidated muds results in extensive extrusion, lateral extension, and diapirism. Resultant contemporaneous structures create a highly irregular slope topography having many closed or partially closed sub-basins (fig. 4). Slope depositional patterns are thus greatly complicated, and transport of sand to the base of the slope becomes difficult. Sub-basin sediments may uplap adjacent diapiric highs.

Key References: Brown and Fisher (1977), Lehner (1969)

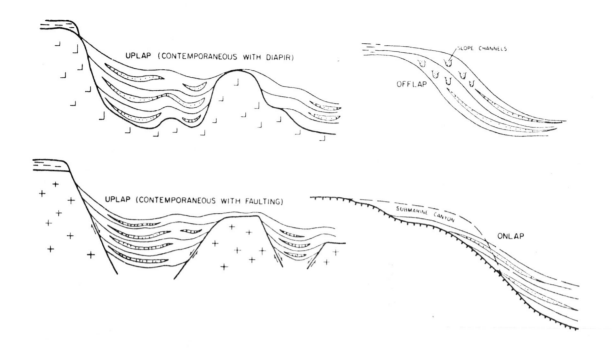

Figure 1. Generalized distribution of sand bodies within the various types of slope systems. From Brown and Fisher (1977).

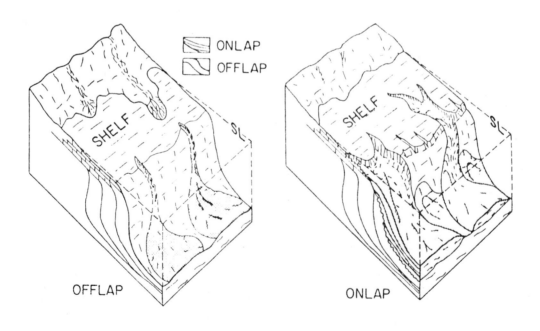

Figure 2. Alternation of offlap (constructional) and onlap (destructional) styles of slope deposition. From Brown and Fisher (1977).

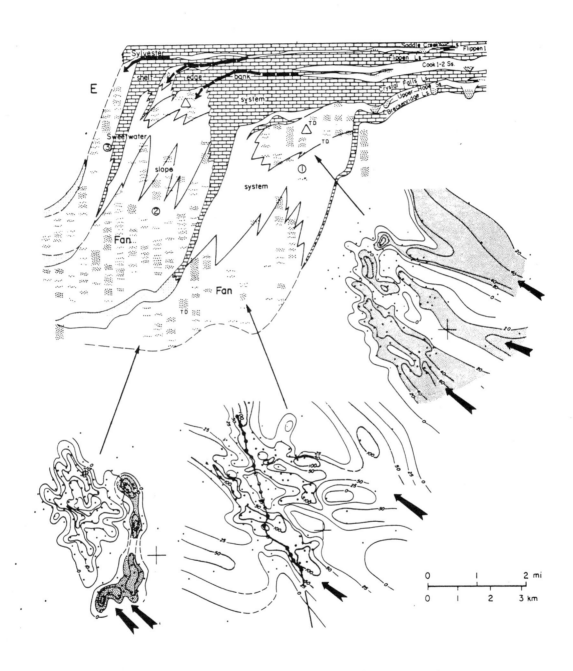

Figure 3. Sand distribution patterns in an offlapping slope wedge (Pennsylvanian/Permian of the Midland Basin). Fan channels give way to irregular suprafan lobes at the base of the slope. Note bimodal sand distribution within the wedge. From Galloway and Hobday (1983); modified from Galloway and Brown (1972).

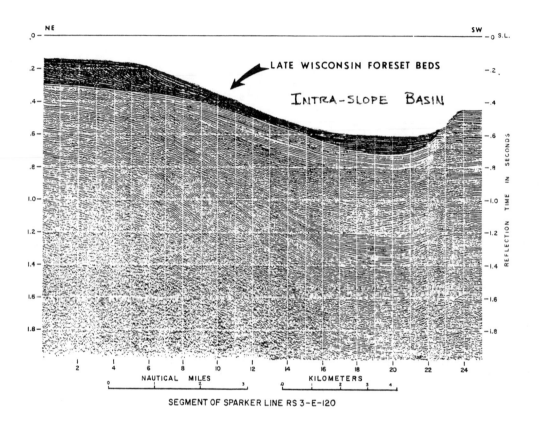

Figure 4. Well-developed Pleistocene clinoforms on the Texas shelf margin. Distal foresets are ponded within an upper slope sub-basin. Thickness of the unit at the shelf edge is about 500 ft (150 m). From Lehner (1969).

REALM OF GRAVITY TECTONICS

The structures that characterize divergent continental margins are almost all examples of gravity tectonics. The spectacular development of diapirism, growth faults, gravity-glide folding, and mass flow along the edges of the Gulf of Mexico qualifies it as the world's type area for currently active gravity tectonics.

The term "gravity tectonics" is more or less self-explanatory: tectonic evolution or movements propelled by the force of gravity. A common misconception is that gravity tectonics is restricted to downward movements because this is the way gravity is generally perceived to operate.

Types of Gravity Tectonics

Figure 1 indicates the range of structures included in the definition of gravity tectonics; two out of three involve at least some upward movements, and the third is dominantly lateral.

1. Gravity gliding: the downslope sliding of rock masses.
2. Gravity spreading: slow, plastic, vertical collapse and complementary lateral spreading (with up and down components) of rock masses.
3. Diapirism: gravitational readjustment of masses with unstable density distribution, such as a density inversion, where denser rock overlies less dense rock. The diapir falls upward under the influence of gravity. Diapirism is used in a loose sense here to include simple doming (pillow formation), which precedes true piercing diapirism.

Size Effects

Gravity influences almost all kinds of geological deformation. Its effects become more pronounced with the size of the structures. For instance, gravity has little influence on folds a few centimeters in size. But in the case of folds with wavelengths measured in kilometers, the pull of gravity greatly shortens the wavelength and amplitude even though the folds may have formed by buckling induced by lateral (nongravitational) forces.

Recognizing Gravity Tectonics

All types of gravity tectonics have two aspects in common. <u>First</u>, the resulting structures must have a geometry showing <u>decreased gravity potential</u> relative to the undeformed state. For example, figure 2 shows a black layer separating a dense overburden from a less dense substratum, which is stippled; a density inversion is therefore present. The large folds formed by gravity tectonics (analogous to salt pillows and shale masses) because the change in gravity potential is negative.

<u>Second</u>, the drop in gravity potential must be large enough to account for the energy dissipated during flow of the rocks, or it must be shown that no other energy besides gravity potential was available during deformation.

All the examples of subaqueous movement in figure 3 are driven by gravity and involve a lowering of gravity potential. <u>Slump and sliding</u> involves gravity sliding at its head and gravity spreading at its toe. <u>Mass flow</u> (which is often confusingly called slumping, a process that should be restricted to rotational movements along curved faults) and <u>turbidity flow</u> are both examples of gravity sliding by viscous flow rather than in discrete units.

<u>Key References</u>: Dott (1963), Ramberg (1981)

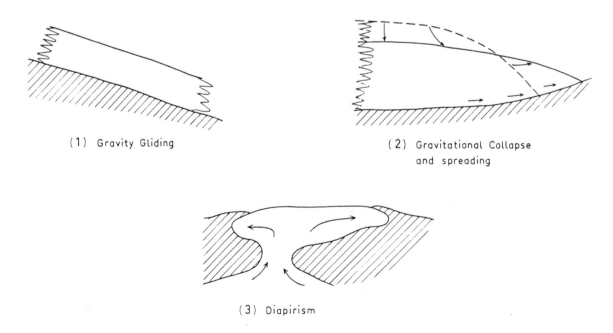

(1) Gravity Gliding

(2) Gravitational Collapse and spreading

(3) Diapirism

Figure 1. Three chief types of gravity tectonics. (From Ramberg, 1981.)

26

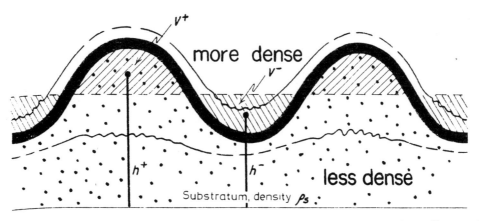

Overburden, density ρ_0

more dense

less dense

Substratum, density ρ_s

Figure 2. Decreased gravity potential as the criterion for gravity tectonics. Change of potential of the large, spontaneously formed folds = $(\rho_s - \rho_0)(h^+ - h^-)V$, where V^+ = volume gained by buoyant source layer = V^- = volume lost by buoyant layer; h^+ and h^- connect centers of gravity of V^+ and V^- to base of system. (Adapted from Ramberg, 1981.)

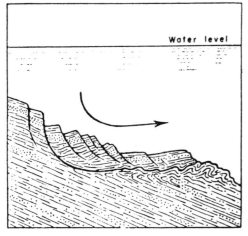

A SUBAQUEOUS SLUMP & SLIDING

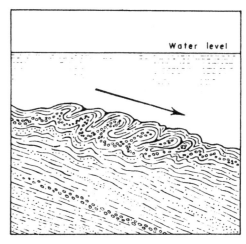

B SUBAQUEOUS MASS FLOW

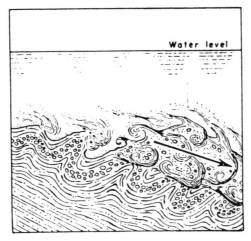

C SUBAQUEOUS TURBIDITY FLOW

Figure 3. Subaqueous gravity movement of sediment: slumping, mass flow, and turbidity flow. (From Dott, 1963.)

27

MODELS OF GROWTH FAULTS

Growth faults, perhaps more usefully known as contemporaneous faults, dominate the structural style of the Texas-Louisiana Gulf Coast. They form contemporaneously with sedimentation, so that their throw increases with depth, and strata on the downthrown side are thicker than those on the upthrown side. Their features are shown in figure 1.

These listric normal faults form in thick, generally regressive, clastic sequences that build out into unconfined depositional sites, such as large deltas that have prograded to the edge of a divergent continental margin. The faults result from contemporaneous failure of the prograding delta slope.

The causes of this failure are still uncertain. Several hypotheses have been advanced, none of which are entirely satisfactory. This suggests that fault processes operate together rather than on their own.

1. Uplift of Salt Ridges or Shale Ridges

The buoyant rise of salt or shale, discussed in unit 14, is one explanation of the cause of growth faulting (fig. 2). Considerable geologic evidence links growth faults and shale ridges on the upthrown sides of the faults.

This model explains the regular spacing of some sets of growth faults as a function of the wavelength of the buoyant ridges that initiated the faults. However, this model alone cannot account for the strong asymmetry of most growth faults, which offset strata predominantly down to the basin. Nor does it account for the large components of horizontal stretching in growth-fault zones. This model proposes that the formation of buoyant ridges triggered the formation of growth faults. But, as will be shown in unit 11, an equally good case can be made that the faults caused ridges of buoyant shale to form underneath them. Spatially the chicken can be seen to be related to the egg, but which came first? Each structure can theoretically form on its own.

2.   Differential Loading

It has been shown experimentally and theoretically that where a dense overburden like sand accumulates with a well-defined front, such as a prograding delta, on a less dense substrate like mud, the difference in loading is greatest at the front of the load. The substrate beneath the front of the load flows laterally away from the load, allowing the load to sink deeper. With sufficient sagging of the load, it yields and is faulted (fig. 3); further sagging takes place by means of these faults because the downthrown sides of the faults trap more heavy sediment, which accentuates local differential loading. Differential loading certainly operates in any growth fault that displaces sand overlying mud. But the model fails to account for the regular spacing of many growth-fault sets, the occurrence of growth faults in all-shale sequences, and the presence of shortening structures far downdip of the sand pinch-out.

3.   Differential Compaction

If two adjacent columns of clastic sediment differ in their proportions of sand and shale because of facies pinch-outs, on compaction each column will reduce in volume by different amounts because of their different properties (fig. 4). This differential compaction causes differential strain on each side of the facies changes. If strain is sufficiently high or rapid, steeply dipping faults form along the facies change. If the total thickness of compacting sediments also varies laterally, as in a seaward thickening wedge, differential compaction (F in fig. 3) will be accentuated by drape compaction (T in fig. 3). In this figure, shale is assumed to compact more than sand. This may be true for the early and late stages of compaction. But the reverse is the case in the middle stages of compaction, when impermeable shales are overpressured. Throughout much of the dewatering compaction history, sand compacts more than shale. Assessing the validity of this model is therefore complex, requiring a knowledge of the burial history and distribution of facies in each case.

Steeply dipping faults that die out downward have been ascribed to differential compaction, but clearly this mechanism cannot account for the highly extensional listric faults that dominate the structural style.

4.   Free Gravity Gliding

This mechanism operates independently of all the others, requiring only a soft rock of low viscosity and a gentle slope of one degree or more. The mechanics of gravity gliding entirely within overpressured shale is described in unit 10. Sliding can also take place where a glide sheet of moderately stiff rock overlies a glide zone of much softer rock, such as evaporites or shale. The slow, solid-state flow of rocks, known as creep, is approximately analogous to the flow of fluids. Thus velocity profiles within the glide zone of ductile rock can be modeled (fig. 5). Velocities decrease downward through the soft layer because of frictional drag along its base. This upward increase is faster in a shear-thinning fluid, which approximates rock behavior better than a Newtonian fluid does. It is obvious from the velocity profiles in figure 5 that the maximum movement of a glide sheet over a glide zone occurs in the case of a stiff rock overlying a soft rock.

Differential velocities can result from differential thicknesses as well. Figure 6 shows how the velocity within a soft glide zone like salt will accelerate as the salt flows over a steplike increase in thickness. Because of the shapes of the velocity profiles in the thin salt and thick salt, this acceleration is fastest close to the base of the thin salt. If this type of glide zone is overlain by a stiff overburden like carbonate or sand, the overburden will be stretched over the step in the glide zone. Growth faults may therefore be initiated in the strata that overlie evaporites early in the history of basin subsidence, later propagating up into younger sediments.

Key Reference: Bruce (1973)

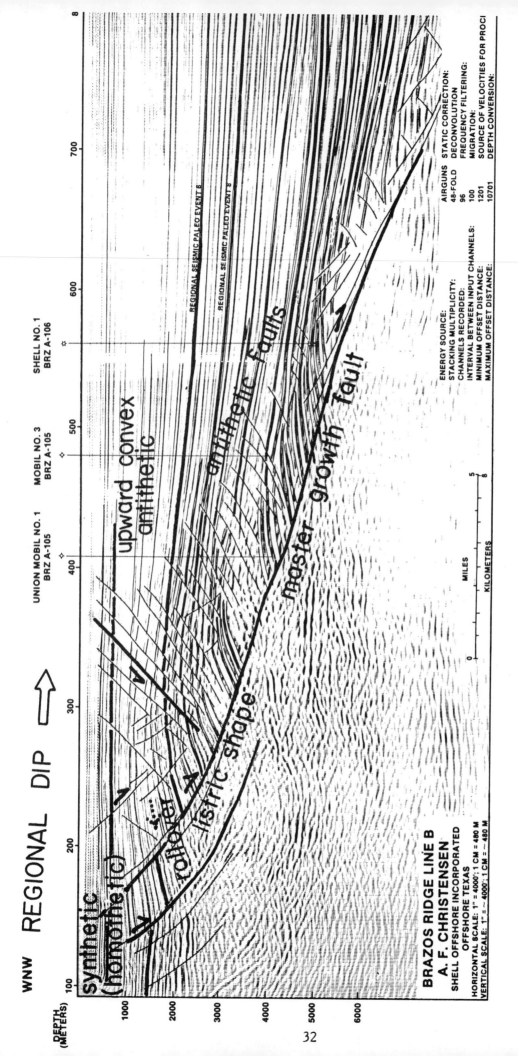

Figure 1. Structure of a Gulf Coast growth fault. (Adapted from Christensen, in Bally, 1983.)

32

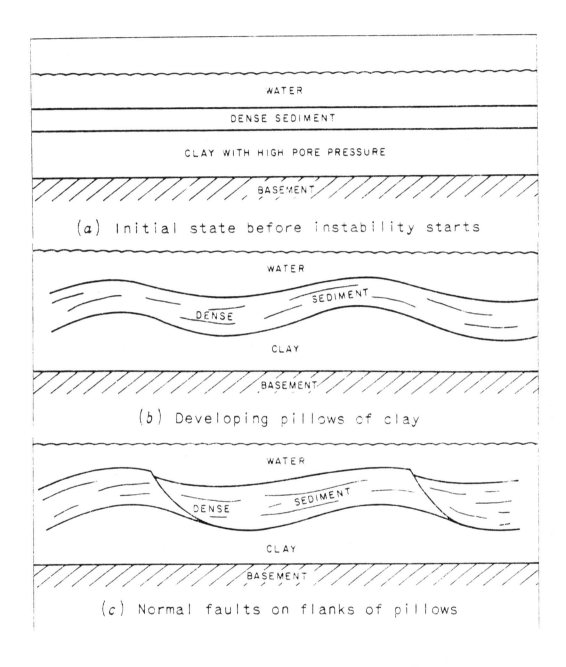

WATER

DENSE SEDIMENT

CLAY WITH HIGH PORE PRESSURE

BASEMENT

(a) Initial state before instability starts

WATER

DENSE SEDIMENT

CLAY

BASEMENT

(b) Developing pillows of clay

WATER

DENSE SEDIMENT

CLAY

BASEMENT

(c) Normal faults on flanks of pillows

Figure 2. Formation of growth faults by uplift of shale ridges. (After Odé, 1962.)

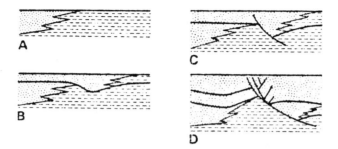

Figure 3. Formation of growth faults due to differential loading of a greater than normal thickness of delta-front sand (stippled) over prodelta mud (dashed). (From Bruce, 1978.)

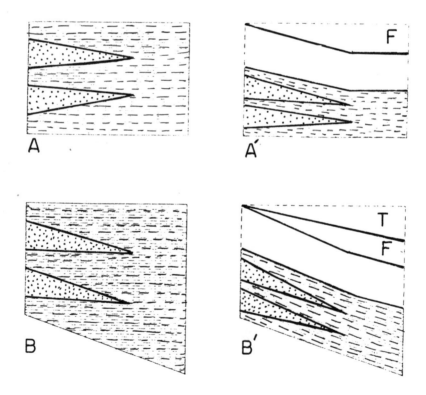

Figure 4. Formation of growth faults by differential compaction due to facies change (A-A') and thickness increase combined with facies change (B-B'). Maximum sand percent at left is 50; during compaction porosity of sand changes from 40 to 20 percent, and shale changes from 80 to 30 percent. F is differential subsidence during compaction due to the facies changes, and T is differential subsidence due to the thickness change. Growth faults would be initiated at point of maximum differential subsidence, which is the front of the sand wedges. (Adapted from Carver, 1968.)

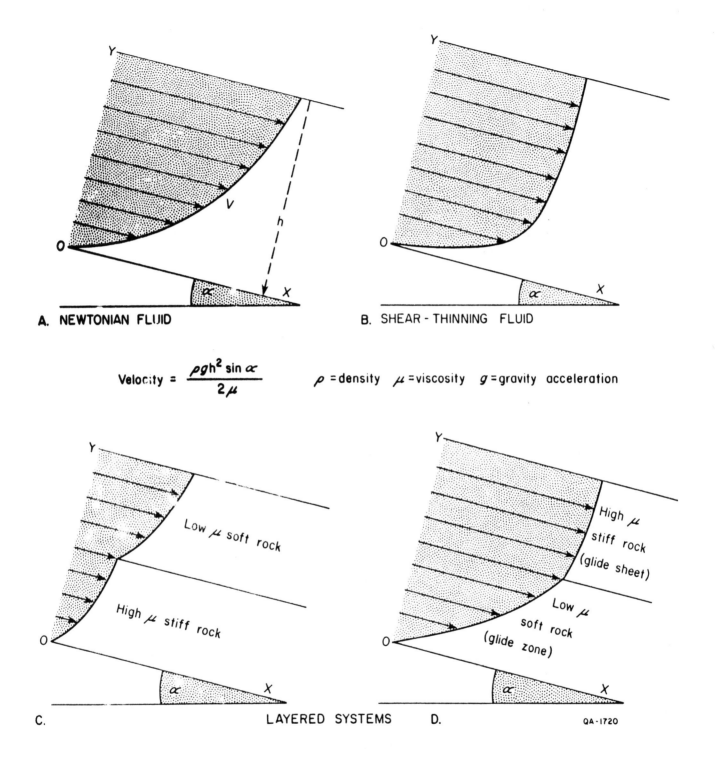

A. NEWTONIAN FLUID

B. SHEAR - THINNING FLUID

$$\text{Velocity} = \frac{\rho g h^2 \sin \alpha}{2\mu}$$

ρ = density μ = viscosity g = gravity acceleration

Low μ soft rock

High μ stiff rock

High μ stiff rock (glide sheet)

Low μ soft rock (glide zone)

C.   LAYERED SYSTEMS   D.   QA-1720

Figure 5. Flow patterns in tilted, tabular bodies of ductile materials. The heavy line connecting the ends of the arrows (velocity vectors) is the velocity profile. A. Velocity profile in true (Newtonian) fluid. B. Velocity profile in a shear-thinning fluid, which is more analogous to flowing rock salt. C. Composite velocity profile in a soft layer overlying a stiff layer. D. Composite velocity profile in a stiff layer overlying a soft layer; velocities are much greater here than in the inverted model in C. (Adapted from Kehle, 1970.)

35

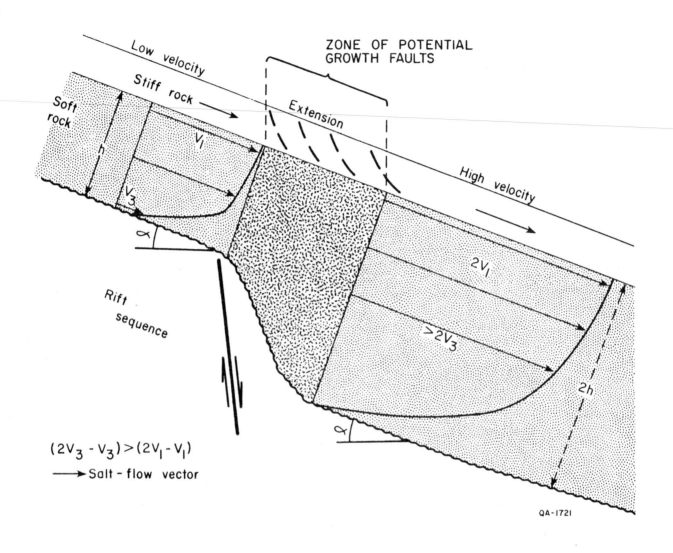

Figure 6. Formation of a zone of extension in a stiff glide sheet overlying a soft glide zone (such as rock salt) of uneven thickness due to deposition across a fault scarp. Acceleration of the flowing glide zone causes normal faulting in the stiff overburden. This model could apply either to salt deposited on a rifted terrane and overlain by carbonate or to uneven thicknesses of shale overlain by sand, typical of the Gulf Coast Tertiary. (Adapted from R. O. Kehle, unpublished data.)

36

THIN-SKINNED GRAVITY SLIDING AS A MECHANISM FOR GROWTH FAULTING

Thin-skinned gravity gliding occurs at relatively shallow depths and is not triggered by deeper deformation. However, deeper deformation can eventually spread upward to deform the growth faults.

Overpressure on the Delta Slope

The head of a prograding delta slope roughly coincides with a facies change from sand-rich deltaic sediments to sand-poor delta-slope sediments (fig. 1). The distal muddy sediments are overpressured (0.46 FPG) because initial compaction reduces their permeability, hindering water expulsion. The model assumes "leaky" overpressure, where fluid continually leaks through the shale into the overlying sand, through which it rapidly escapes (fig. 2). Here the subvertical effective stress, σ_{zz}, remains constant within the overpressured zone where the fluid gradient, II, is the same as the lithostatic overburden gradient, σ_{zz}. In the case of a "sealed" overpressured zone, the fluid pressure increases abruptly downward across the sealing strata (fig. 2).

On a stable slope with normal fluid pressure, the system is in static equilibrium. The downslope component of the weight of the fluid-filled loose sediment is balanced by the reactive upslope shear stress, τ, which results from the shear strength of the sediment caused by friction at particle contacts. If pore pressure remains hydrostatic (balancing the full weight of the water column only) with increasing burial, the grain skeleton carries the full weight of the increasing sediment thickness by interparticle contact. The shear strength thus increases with depth, like the pore pressure.

But once pore pressure begins to increase faster than the hydrostatic increase because of impermeability, instability is induced by overpressuring as shallow as 10 m (30 ft) under deltaic conditions. As the overpressure increases with depth, the excess pore pressure acts to separate the grains and weaken the rock because with less friction between the grains, the rock has lower shear strength. The effective normal stress, σ'_1, (total normal stress minus pore pressure), declines downward to a critical depth, Z, where the effective normal stress is too weak to hold the sediment together against the shear stress caused by the delta slope. The sediment begins to shear at point A (fig. 3). Overlying strata begin sliding along a basal slip plane, which propagates downdip as a zone of shear-weakened strata at depth Z (table 1).

Formation of the Glide Sheet

On an infinite slope the glide sheet extends at its head by "active" gravity gliding, and shortens at its toe by "passive" gravity spreading. Lateral stresses are transmitted from head to toe by means of an intervening stiff part of the glide sheet in this model. In simple terms, the base of the glide sheet is a listric normal fault at its head, a bedding-parallel fault along its midsection, and a thrust fault at its toe.

To understand why this characteristic shape develops, consider the slip-line field (fig. 4). Slip lines are the traces of theoretical faults calculated from the known stresses. Not all these slip lines will be activated as faults. But for a given stress regime, faults theoretically form only along the slip lines calculated for that particular regime. Two sets of slip lines intersect in conjugate fashion, their acute angle of intersection being bisected by the maximum principal compressive stress, σ_1.

In hydropressured strata (fig. 4, top) the slip lines define the familiar conjugate pattern for planar normal faults in the "active" head and planar reverse faults in the "passive" toe. In strata overpressured below the depth Z_{ip} (fig. 4, bottom) all faults curve because the stress regime changes with depth as overpressure builds. Near the top, σ_1 is near-vertical, but as the effective normal stress declines with increasing depth, σ_1 curves to an angle of about 30° with the basal slip plane, which is parallel to the gentle surface slope. Figure 5 shows a correlation between the shapes of early to middle Tertiary growth faults and the present-day top of the geopressured zone.

During deformation, synthetic faults are activated parallel to slip lines marked β and antithetic faults form parallel to slip lines marked α. The antithetic faults accommodate the rotation of strata as they slide over a curved basal slip plane. Note that the antithetic faults are convex upward (unlike the synthetic master fault, which is concave upward) in contrast to the shape that they are often given in seismic sections or geologic cross sections. This convex upward shape is required because of the curve of σ_1. Furthermore, a simple experiment with scissors and paper will show that only convex upward antithetic normal faults can result in rollover (fig. 6).

A slide on an infinite slope that is not buttressed at its lower end is known as open-ended; reverse faults are absent because the structure has no toe. Each family of growth faults passes downward into the same basal slip plane. In this model each family is regularly spaced from the next youngest family downdip. The spacing is dependent on the thickness of the sliding unit and its dip. In nature this spacing is less regular because of numerous irregularities in the strata, and the influence of other forces.

A _complete_ slide structure has a toe region where strata are shortened (fig. 4). This structure forms when the rate of sedimentation is high compared with the rate of overpressure relaxation.

Internal Geometry of the Slide Sheet

The geometry of strata within the sliding sheet downdip of the growth fault is controlled by three main variables: _rate of sedimentation_, _slide velocity_, and _velocity profile_. Where the rate of sedimentation is greater than the slide velocity, rollover is weak (fig. 7A). Where the rate of sedimentation is less than the slide velocity, rollover is more pronounced and more effective as a trap (fig. 7B). Variations in slide-velocity profiles can be caused by the variation in lateral compactibility of strata in the midsection of a slide sheet. _Overflow_ (fig. 7C) is the most typical profile, resulting from frictional drag along the base of the slide; the resulting geometry is typical of the style of Gulf Coast growth faults. If higher compactibility of deeper units causes _underflow_ (fig. 7D), the lazy-Z-shaped geometry of the strata is quite different from a simple rollover (fig. 8). Analogous differences arise from overflow and underflow in the toe of the slide (fig. 7E, F). Here antithetic thrust faults are encouraged by overflow.

All these computer-generated cross sections of the internal structure are based on the final velocity profiles shown. However, a more realistic model is one that is built incrementally using successive parts of the profile (fig. 7G). The final master growth fault envelopes the older synthetic faults, which have migrated updip by serial formation. This migration has caused the rollover of the lower, older faults to be _reversed_ by many small fault displacements. A seismic interpreter might incorrectly assume that this poorly defined zone of rollover reversal is due to drag on the master fault instead of displacement by many small synthetic faults.

Overpressure Relaxation Structures

Postdepositional strain can result from relaxation of the overpressure that caused the growth faulting. Whether the overpressured zone leaks upward to the surface or downward along the basal slip plane, the faults that form successively become straighter as the overpressure subsides (fig. 9). If the succession of faults radiates from a common point, horsetails can form.

Key References: Crans and others (1980), Mandl and Crans (1981)

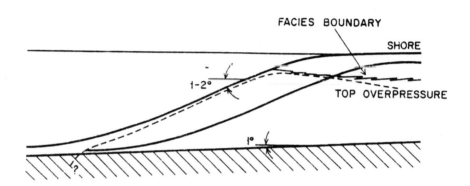

Figure 1. Idealized dip section of a prograding delta slope. (From Mandl and Crans, 1981.)

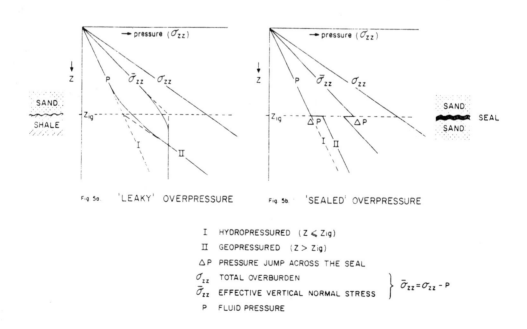

Figure 2. Pressure-depth diagrams depicting two idealized types of overpressure in shale overlain by hydropressured sand. (From Crans and others, 1980.)

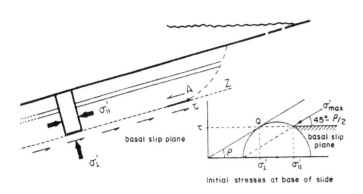

Figure 3. A slope on the verge of gliding; shear stress, τ, at the base of the slide at depth Z is shown by Mohr diagram. (From Mandl and Crans, 1981.)

λ	Critical Slope for Gravitational Sliding Degrees
0.46	17.2
0.60	13.0
0.70	9.8
0.80	6.6
0.85	5.0
0.90	3.3
0.94	2.0
0.97	1.0

Table 1. Critical inclination necessary for gravity sliding (after Rubey and Hubbert, 1959).

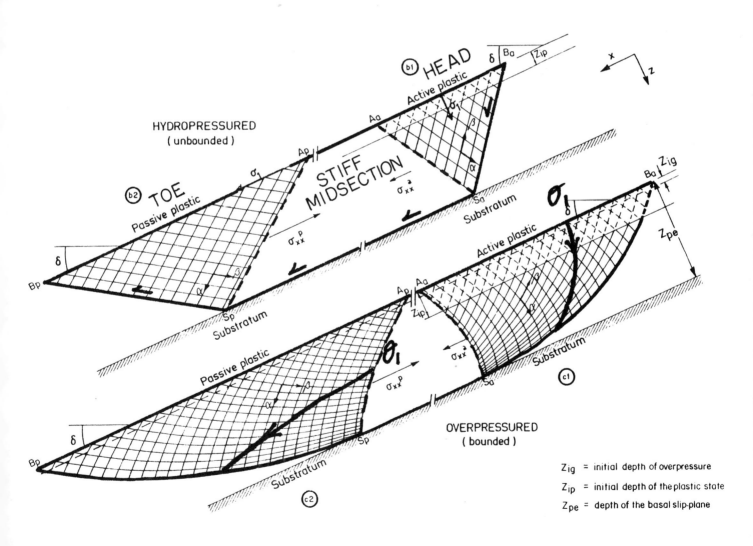

Figure 4. Potential fault patterns (slip-line fields) for a glide sheet composed of hydro-pressured and overpressured shale. (Adapted from Crans and others, 1980.)

41

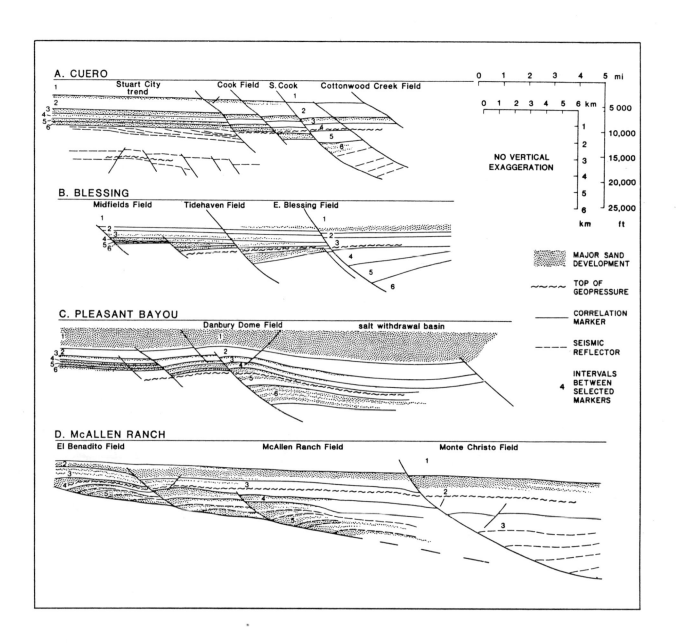

Figure 5. Dip sections through Texas Gulf Coast Tertiary strata showing how planar growth faults commonly become listric below the present top of geopressure. This suggests that the geopressured zone has remained at roughly constant depth since the formation of the faults. A. Wilcox Group; B. and C. Frio Formation; D. Vicksburg Group. (From Winker and others, 1983.)

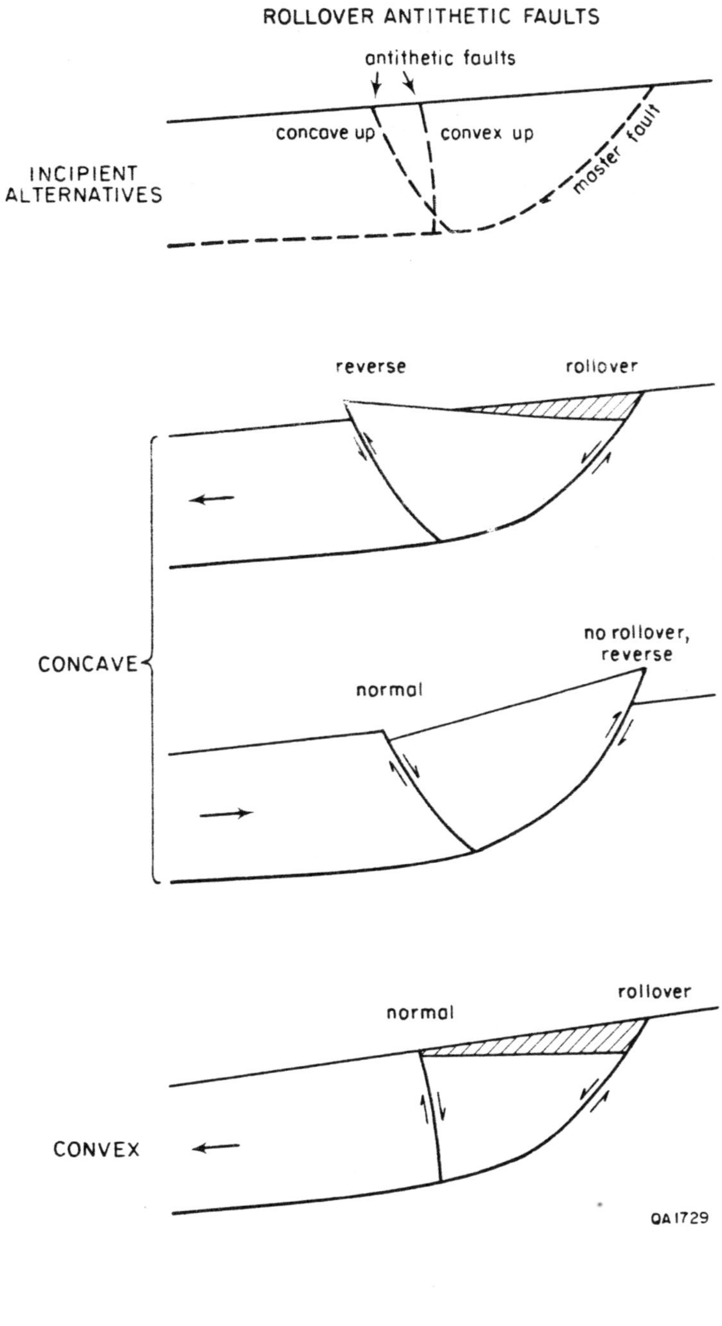

Figure 6. Schematic cross sections of antithetic faults associated with listric normal faults. Only an upward convex normal antithetic fault can produce rollover against a listric normal master fault by faulting alone, in agreement with the theoretical antithetic faults in figure 4. Rollover can also be produced by flexural-slip folding (analogous to the bending of a telephone book and illustrated in unit 21, figure 6).

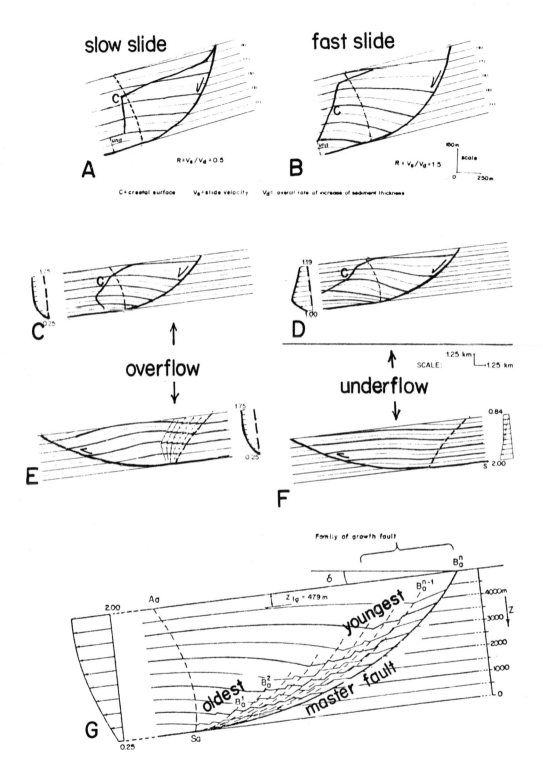

Figure 7. Internal structure of numerically modeled glide sheets. A. Slight rollover in the head where sediment-accumulation velocity is greater than slide velocity; c is crestal trace, dashed line is potential antithetic fault. B. Pronounced rollover in the head where slide velocity is greater than sediment-accumulation velocity. C. Single curve of bedding in the head where upper part of glide sheet slides faster than lower part (overflow). D. Double curve (lazy Z) of bedding in the head where lower part of glide sheet slides faster than upper part (underflow) (see fig. 8). E. Pronounced anticline in the toe produced by overflow. F. Slight anticline in the toe produced by underflow. G. Synthetic faults and complex rollover structure in the head modeled by assuming syndepositional rise of the top of leaky overpressure. Synthetic faults migrate updip with time and produce the appearance of pseudonormal drag on the downthrown side of the master fault. (Adapted from Crans and others, 1980.)

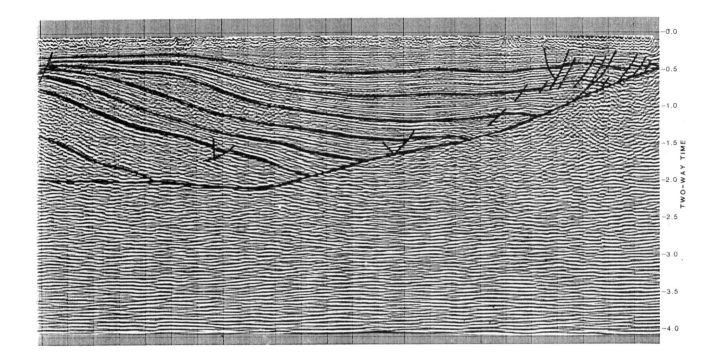

Figure 8. A geometric (not geologic) analog of figure 7D in the form of a Miocene listric normal fault in the Marys River Valley, Basin and Range Province, Nevada. (From Robison, in Bally, 1983.)

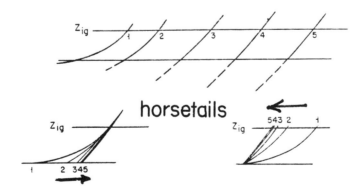

Figure 9. Postdepositional horsetail faults theoretically produced during decline of overpressure; depth to top of overpressure (Z_{ig}) is constant. By the time fault 5 forms, the strata are virtually hydropressured, so the fault is almost planar. (Adapted from Mandl and Crans, 1981.)

INTERACTIONS BETWEEN GROWTH FAULTS AND SEDIMENTATION

The previous section described how overpressured shales could slide down extremely gentle slopes under their own weight, initiating listric normal faults at the updip end of the gliding mass. In a prograding margin such shales are subsequently buried beneath coarser regressive clastics. This progradational burial initiates several types of deformation that complicate analysis of the continental margin.

Sand-Shale Spoon

Sedimentary loading during progradation flexes the edge of the sedimentary basin; maximum subsidence is below the current depocenter. Figure 1 shows the resulting spoon shape in cross section (with great vertical exaggeration), comprising two half-spoons. The basal shale-dominated wedge is overpressured, initially buoyant, and of low viscosity. The overlying sand-dominated wedge compacts readily during burial so it is hydropressured, initially denser, and more viscous than the underlying shale. The compactible half-spoon of sand wedges out seaward of the shelf break, which overlies the depocenter. Here the noncompactible half-spoon of shale is thickest. The megafacies line separating the two half-spoons dips at about 2° landward.

Gravity Response to the Spoon

At the front of the sand wedge on the megafacies line the setting is ripe for gravity tectonics. The heavier sands sink into the initially buoyant shales because of the density inversion. Because the sands wedge out seaward, the displaced, mobile shales flow in this direction. With enough rotational sag the sandy unit breaks on listric normal faults (fig. 2, box 2). This sag allows more sand to accumulate on the downthrown side; more shale flows laterally. After enough rotational faulting, a shale mound forms seaward of the growth fault (box 6). The mound acts as a focus for the next growth fault (box 7). Growth faults form serially upward and seaward along the megafacies line as the delta progrades and aggrades (fig. 3). In all growth faults the displacement along a single fault increases downward, but the maximum rate of displacement, as measured by the expansion index (ratio of sediment thickness on the downthrown side to sediment thickness on the upthrown side), commonly occurs in the middle of the fault (fig. 4). Plotting the expansion index differentiates a syndepositional listric normal fault (growth fault) from a postdepositional listric normal fault. After accounting for the effects of compaction, the former has an index of more than 1, whereas the latter has an index of 1.

During burial the sands compact faster than the underlying, more distal shales. Thus, strata within the increasingly dense compactible half-spoon of sand subside faster than those in the uncompactible half-spoon across the megafacies line. Accordingly, the line rotates so that its lower proximal end subsides many thousands of feet in the deepest part of the spoon (fig. 5A). This rotational sagging of the megafacies line squeezes mobile, overpressured shale seaward on a scale far greater than that of individual growth faults. The shale mounds on the upthrown, landward side of each growth fault rise diapirically through the thickening, advancing wedge of sand, or they spread laterally by asymmetric folds and thrusts verging seaward at the toe of the shale (fig. 5B).

This enormous mass transfer of shale can profoundly alter the primary depositional geometry. In figure 6, a 3,000-m (10,000-ft) thick shale mound has formed by seaward flow of shale ahead of a series of growth faults. The contact between A and A' is probably the basal slide into which the left-hand faults sole out. The mounded shale may have originally onlapped this contact in the undeformed state. The B-B' contact is probably the folded megafacies boundary.

Key References: Dailly (1976), Winker and Edwards (1983)

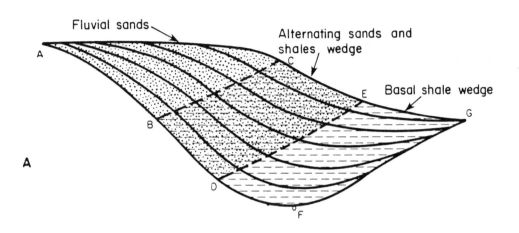

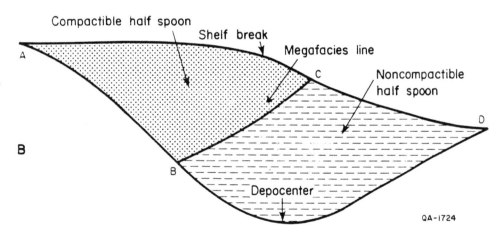

Figure 1. The regressive megasequence spoon. (Adapted from Dailly, 1976.)

49

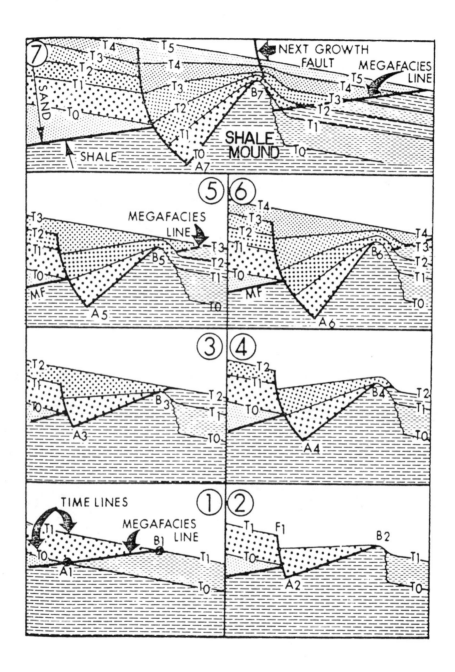

Figure 2. The birth and death of a growth fault and coeval shale mound. B_1 is the leading edge of the sand half-spoon. (Adapted from Dailly, 1976).

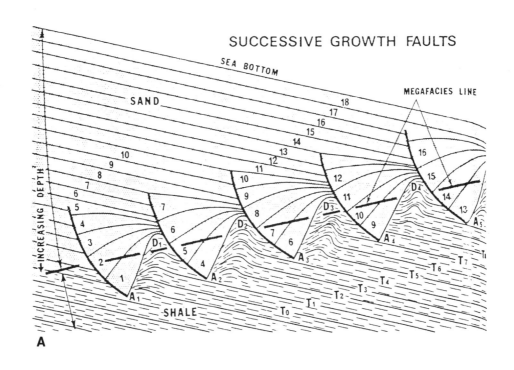

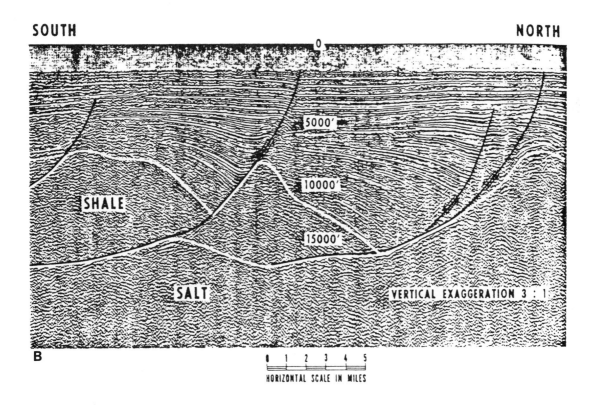

Figure 3. A. Formation of a series of growth faults (A_1 to A_5) along the megafacies line by the mechanism shown in figure 2. (From Dailly, 1976.) B. Analogous growth faults and shale mounds below the outer continental shelf of the Gulf of Mexico. (From Woodbury and others, 1973.)

PLOT OF EXPANSION INDICES

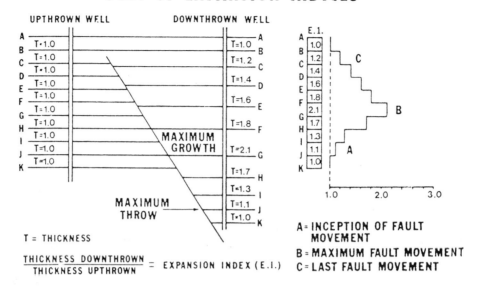

T = THICKNESS

$$\frac{\text{THICKNESS DOWNTHROWN}}{\text{THICKNESS UPTHROWN}} = \text{EXPANSION INDEX (E.I.)}$$

A = INCEPTION OF FAULT MOVEMENT

B = MAXIMUM FAULT MOVEMENT

C = LAST FAULT MOVEMENT

Figure 4. Expansion index as a measure of the timing and rate of movement along a growth fault. Although displacement increases downward to horizon K, the rate of displacement reaches a maximum at horizons F and G. (From Dailly, 1976, modified from Thoren, 1963.)

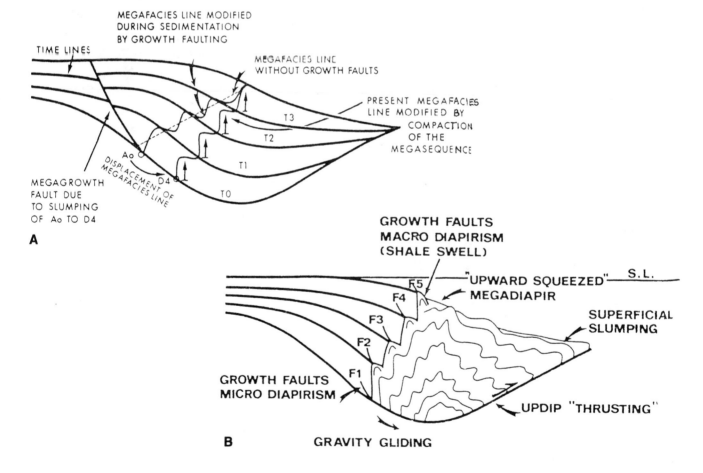

Figure 5. A. Migration of the regressive megafacies line owing to greater compaction of the sand half-spoon updip relative to the shale half-spoon downdip. B. Structural effects of this migration. (From Dailly, 1976.)

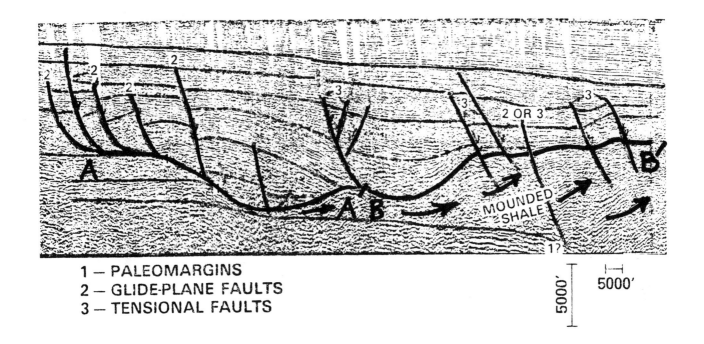

1 — PALEOMARGINS
2 — GLIDE-PLANE FAULTS
3 — TENSIONAL FAULTS

Figure 6. A possible natural analog of the model in figure 5. (Adapted from Yorston and Weisser, 1982.)

GROWTH FAULTS AND PETROLEUM TRAPS

In both the Gulf Coast and Niger Delta, hydrocarbons tend to accumulate on the downthrown sides of the growth faults (fig. 1). Trapping is controlled by rollover of regional dip, fault-related facies changes, and sealing faults.

Rollover Anticline

Reversal of regional dip by rotation of the downthrown block along a listric normal fault forms the classic rollover anticline, whose crest is centered opposite that part of the spoon-shaped fault having the greatest throw (fig. 2). Hydrocarbons are trapped in this culmination, downdip of the actual fault trace (figs. 2 and 3A).

Sand Pinch-outs Down Regional Dip

The sands concentrated downdip of each growth fault wedge out seaward. In zones of significant rollover (fig. 3B) this pinch-out is up the local dip, creating a combined structural-stratigraphic trap.

Sealing Faults and Migration Paths

Faults, themselves, can trap hydrocarbons if the faults are of the sealing type (fig. 4). At deep levels, rotation by listric faulting forms structures with their highest points against the upthrown side of the next growth fault seaward (fig. 3C). In some cases, as in the Akata Formation of the Niger Delta, the fault intersection with the upper bedding plane of the reservoir is the spill point of the accumulation (fig. 5). These spill points are thought to be the entry points for hydrocarbons from the fault zone into the reservoir. In the Niger Delta the growth faults act as channelways for vertical migration if the throw is larger than 150 m (500 ft) (fig. 5). Smearing of sand and shale into the fault plane gouge by intense simple shear creates strong anisotropy and enhances permeability along the fault zone. This anisotropy also acts as a sealing mechanism by preventing flow across the fault zone if the zone has been passed on the downthrown side by an interval consisting of more than 25 percent shale.

Key Reference: Weber and Daukoru (1975)

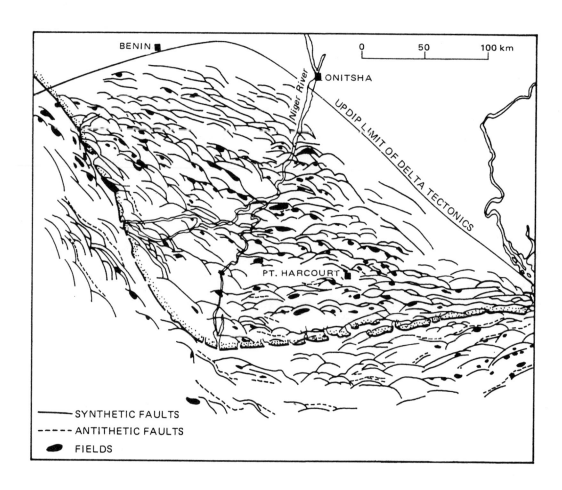

Figure 1. Growth faults and hydrocarbon accumulations in the Niger Delta. (From Weber and Daukoru, 1975.)

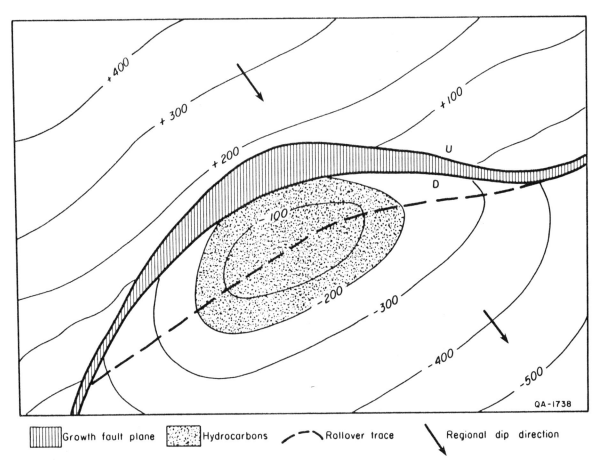

Growth fault plane   Hydrocarbons   Rollover trace   Regional dip direction

Figure 2. Hydrocarbon trap formed by rollover on the downthrown side of a growth fault. (Adapted from Silver, 1979.)

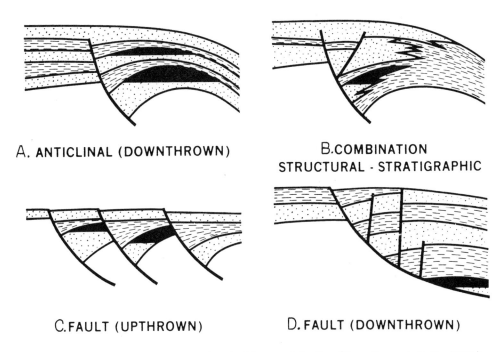

A. ANTICLINAL (DOWNTHROWN)

B. COMBINATION STRUCTURAL - STRATIGRAPHIC

C. FAULT (UPTHROWN)

D. FAULT (DOWNTHROWN)

Figure 3. Hydrocarbon traps associated with growth faults. (From Bruce, in Bally, 1983.)

HYPOTHETICAL SITUATION	ANALYSIS OF FAULT SEAL	
	VERTICAL MIGRATION	LATERAL MIGRATION
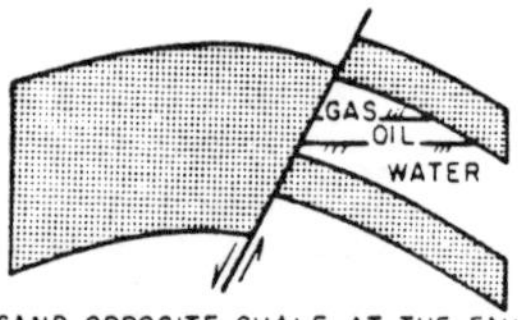 (a) SAND OPPOSITE SHALE AT THE FAULT. HYDROCARBONS JUXTAPOSED WITH SHALE.	SEALING	SEALING RESERVOIR BOUNDARY MATERIAL MAY BE THE SHALE FORMATION OR FAULT ZONE MATERIAL.
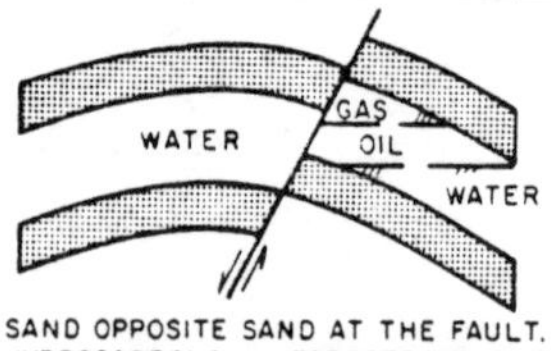 (b) SAND OPPOSITE SAND AT THE FAULT. HYDROCARBONS JUXTAPOSED WITH WATER.	SEALING	SEALING SEAL MAY BE DUE TO A DIFFERENCE IN DISPLACEMENT PRESSURES OF THE SANDS OR TO FAULT ZONE MATERIAL WITH A DISPLACEMENT PRESSURE GREATER THAN THAT OF THE SANDS.
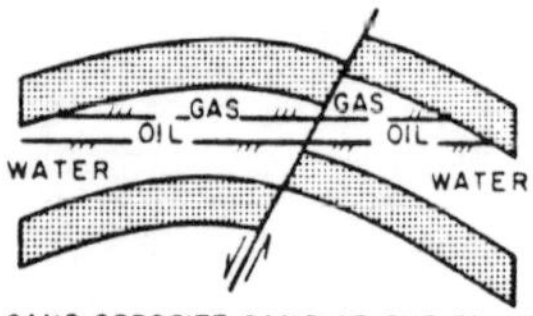 (c) SAND OPPOSITE SAND AT THE FAULT. COMMON HYDROCARBON CONTENT AND CONTACTS.	SEALING	NONSEALING POSSIBILITY IS REMOTE THAT FAULT IS SEALING AND THE RESERVOIRS OF DIFFERENT CAPACITY HAVE BEEN FILLED TO EXACTLY THE SAME LEVEL BY MIGRATING HYDROCARBONS.
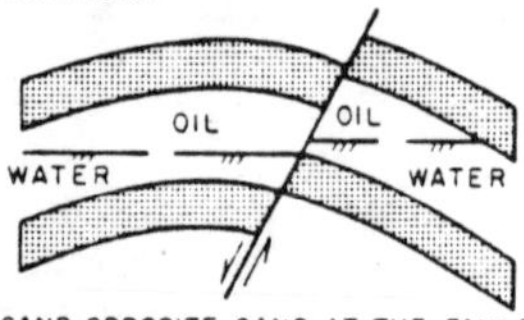 (d) SAND OPPOSITE SAND AT THE FAULT. DIFFERENT WATER LEVELS.	SEALING	UNKNOWN NONSEALING IF WATER LEVEL DIFFERENCE IS DUE TO DIFFERENCES IN CAPILLARY PROPERTIES OF THE JUXTAPOSED SANDS. SEALING IF WATER LEVEL DIFFERENCE IS NOT DUE TO DIFFERENCES IN CAPILLARY PROPERTIES OF THE JUXTAPOSED SANDS.
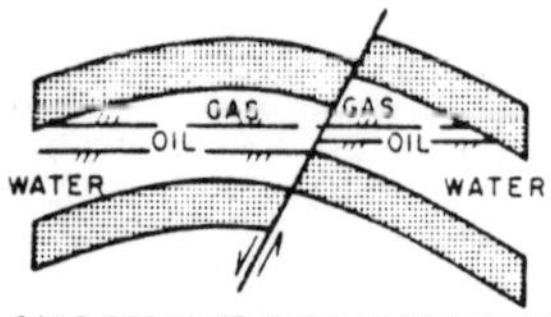 (e) SAND OPPOSITE SAND AT THE FAULT. COMMON GAS-OIL CONTACT, DIFFERENT OIL-WATER CONTACT.	SEALING	NONSEALING POSSIBILITY IS REMOTE THAT FAULT IS SEALING AND MIGRATING GAS HAS FILLED THE RESERVOIRS OF DIFFERENT CAPACITY TO EXACTLY THE SAME LEVEL.
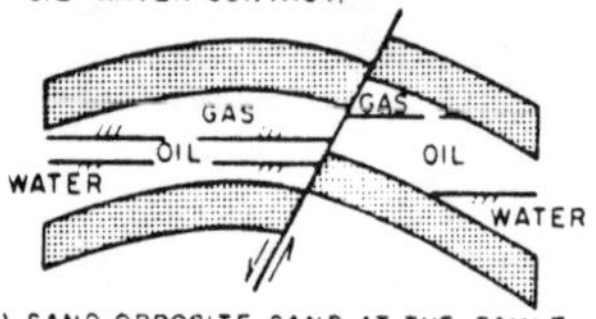 (f) SAND OPPOSITE SAND AT THE FAULT. DIFFERENT GAS-OIL AND OIL-WATER CONTACTS.	SEALING	SEALING A DIFFERENCE IN BOTH GAS-OIL CONTACT AND OIL-WATER CONTACT INFERS THE PRESENCE OF BOUNDARY FAULT ZONE MATERIAL ALONG THE FAULT.
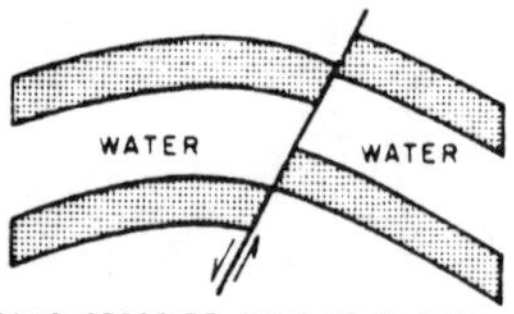 (g) SAND OPPOSITE SAND AT THE FAULT. WATER JUXTAPOSED WITH WATER.	UNKNOWN	UNKNOWN

Figure 4. Sealing and non-sealing conditions across a normal fault. (From Smith, 1980.)

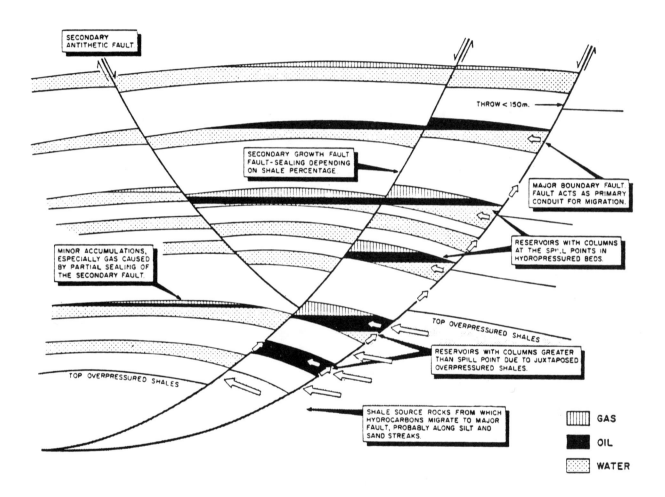

SECONDARY
ANTITHETIC FAULT

THROW < 150m.

SECONDARY GROWTH FAULT
FAULT-SEALING DEPENDING
ON SHALE PERCENTAGE

MAJOR BOUNDARY FAULT.
FAULT ACTS AS PRIMARY
CONDUIT FOR MIGRATION.

RESERVOIRS WITH COLUMNS
AT THE SPILL POINTS IN
HYDROPRESSURED BEDS.

MINOR ACCUMULATIONS,
ESPECIALLY GAS CAUSED
BY PARTIAL SEALING OF
THE SECONDARY FAULT.

TOP OVERPRESSURED SHALES

TOP OVERPRESSURED SHALES

RESERVOIRS WITH COLUMNS GREATER
THAN SPILL POINT DUE TO JUXTAPOSED
OVERPRESSURED SHALES.

SHALE SOURCE ROCKS FROM WHICH
HYDROCARBONS MIGRATE TO MAJOR
FAULT, PROBABLY ALONG SILT AND
SAND STREAKS.

▥ GAS

█ OIL

▦ WATER

Figure 5. Schematic dip section across a field in the Niger Delta showing relations between fault geometry, top of overpressure, and hydrocarbon traps. (From Weber and Daukoru, 1975, after R. G. Precious.)

SALT STRUCTURES

<u>Key Reference</u>: Trusheim (1960)

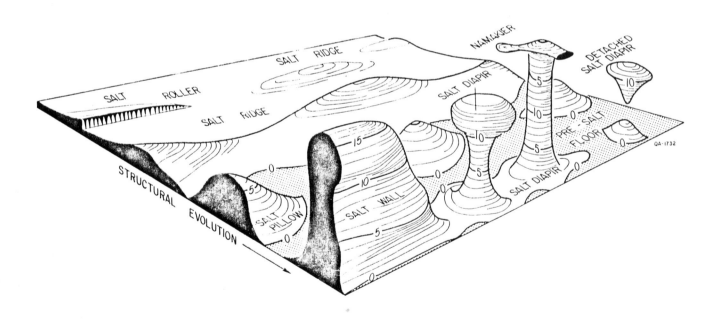

Figure 1. The main types of large salt structure. Structure contours are in arbitrary units. Salt nappes of the Sigsbee Scarp type are not shown because they are an order of magnitude larger.

HOW SALT AND SHALE FLOW

To understand diapirism, we must first establish why certain sedimentary rocks flow and others do not. So we examine the <u>rheology</u>, or flow properties of these rocks. The most common diapirs in sedimentary sequences consist of salt, shale, or peat. Peat diapirs are invariably small and are inconsequential on the Gulf Coast, so we can ignore them.

Flow of Salt

Thanks to intensive study of the deformational properties of rock salt for the National Waste Terminal Storage program, our knowledge of salt rheology has greatly expanded in recent years. Because we are interested in the behavior of rocks, we look at the flow of polycrystalline halite aggregates rather than single crystals.

Figure 1 shows recent experimental data on dry salt. The dashed lines refer to temperature versus depth and show the present-day extremes of the geothermal gradient in the Gulf Coast. But during the Cretaceous, when salt was active, the gradient would have been even higher than 36° C/km (2° F/100 ft). A rise in temperature encourages ductile flow, other factors being equal. The estimated limits of steady-state stress differences (Ws) are 0.5 and 1.5 MPa (5 and 15 bars) in the upper parts of salt stocks, as calculated from subgrain sizes. Natural steady flow at the present depth of sampled salt stocks is therefore driven by stresses within the limits of these curves. At deeper levels Ws is as high as 20 MPa. The solid lines refer to temperature versus strain rate ($-\log \dot{e}$/s). A value of 15 on the top abscissa signifies that a rock elongates at a rate of 10^{-15} of its length per second, a very slow strain rate. A value of 9 is a million times faster.

This graph shows that dry salt can deform at much lower temperatures (and therefore at shallower depths) than was previously thought possible. The average growth rate of salt diapirs in East Texas has been estimated at 10^{-15}/s. With the minimum stress difference of 0.5 MPa (5 bars), salt could flow at this rate at 57°C (135°F), 900 m (3,000 ft) deep. With a stress difference of 1.0 MPa (10 bars), dry salt can flow at this rate at 25° C (75° F), corresponding to zero burial.

But slightly damp salt can flow even faster. Salt glaciers (namakiers) in the arid Iranian Zagros Mountains flow fastest after a brief shower. Rock salt is about 100 times less viscous (more fluid) at humidities greater than 75 percent than it is in dry air. Strain rates for dampened salt glaciers have been measured at 10^{-9} to 10^{-11}/s. Addition of water has strain-softening effects similar to raising the temperature. In the subsurface a halite evaporite bed is commonly exposed to unsaturated water during compaction of the clastic rocks above and below it. Factors like thickness variations, strength, or density of the surrounding clastic rocks control whether diapirism occurs. Nevertheless, there is no minimum depth required for salt to flow at the slow rates typical of diapirism.

Flow of Shale

Unlike salt, which can flow throughout its burial history, shale requires more specific conditions to flow. On rapid burial, shale gradually compacts. But because of impermeability, pore water is expelled much more slowly than from the surrounding sands. By keeping the grains apart, the excess pore pressure counteracts the lithostatic (geostatic) load pressure and lowers the effective stress (effective stress equals lithostatic stress minus the pore pressure). This weakens the rock and promotes flow.

Pore-pressure profiles based on measurements in shale are similar to profiles calculated by mathematically consolidating shale. This suggests that compaction is the primary cause of overpressure in shales although other processes contribute (such as aquathermal pressuring, water expulsion by the smectite-to-illite conversion, methane gas generation, osmosis, structural barriers to lateral migration like faults, influx of juvenile water, mineral dissolution or precipitation, and tectonic strain). Thus, changes in pore pressure and density can be modeled with reasonable accuracy on the basis of compaction only.

Figure 2A shows the increases in density and pore pressure in shale compacting only under its own weight at time t_0. Overpressuring extends virtually to the surface, and both density and pore-pressure curves are convex upward. Figure 2B shows burial of the shale by a permeable hydropressured overburden (like sand) at subsequent times t_1 and t_2. Because the shale dewaters most easily at its top contact with the permeable overburden, compaction is fastest here, producing the S-shaped kink near the top of the density curves.

Relations between Shale Density and Overpressure

The general trend with increasing burial and compaction of the shale is for the bulk of the shale section to be increasingly overpressured (fig. 2B). But although the general trend is also toward increasing density with time and depth, the profile of each density curve becomes increasingly steep with time. If the sandy overburden has a mean density equivalent to the dashed line, the upper one third of the shale mass at time t_0 is less dense than the overburden; this density inversion is a prime cause of diapirism. By time t_1 a much smaller proportion of the shale is less dense than the overburden. By time t_2 the density inversion has vanished altogether, although the shale is still overpressured (fig. 3).

Although overpressure can persist almost indefinitely in thick shale below permeable sediments, a density inversion exists only during the early stages of burial. Thus, shale can only rise diapirically soon after burial (unlike salt). But in cases of differential loading, such as a sandy delta prograding over marine shales, even without a density inversion the overpressured shales are sufficiently weak (under low effective stress) that they can be squeezed laterally and form mudlump-type diapirs at the delta front, as described in unit 11.

Key References: Bishop (1976), Carter and Hansen (1983)

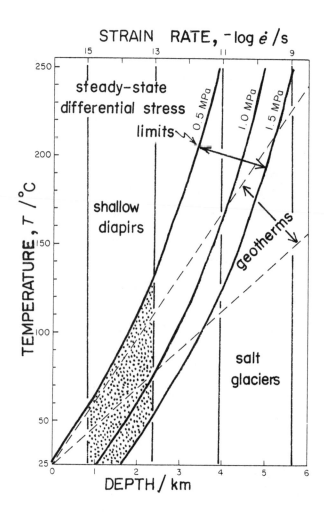

Figure 1. Experimentally deduced flow properties of dry rock salt (solid lines) under varying conditions of temperature, strain rate, and differential stress. Dashed lines show present-day extremes of geothermal gradient in the Gulf of Mexico. (Adapted from Carter and Hansen, 1983.)

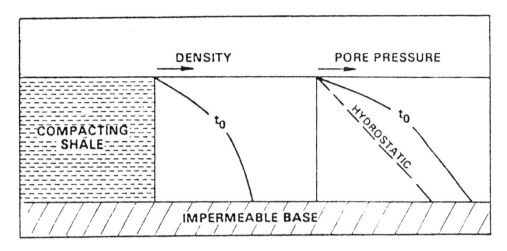

(a) Before burial: The shale mass described in terms of density and pore pressure at t_0, which is the time marking the end of its accumulation and prior to its burial.

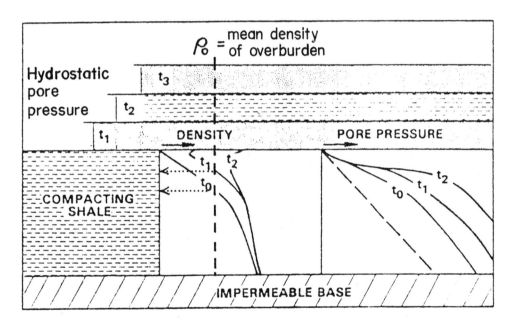

(b) During burial: Evolution of the shale mass during increments of loading.

Figure 2. Compaction-induced changes in pore pressure and density of shale (A) before and (B) after burial beneath hydropressured permeable sand at times t_0 through t_3. (Adapted from Bishop, 1976.)

67

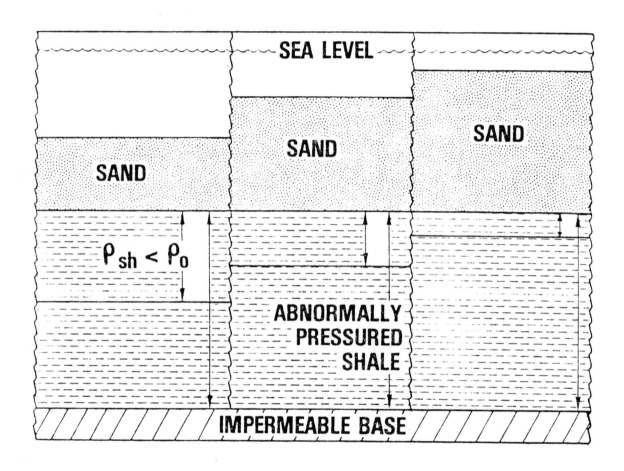

Figure 3. Evolution of shale density inversions (from left to right) as shale is progressively buried. The density inversion shrinks according to the model in figure 2 and will disappear with sufficient burial. (From Bishop, 1976.)

INITIATION OF UPWARD-MOVING GRAVITY STRUCTURES

True (Newtonian) liquids will rise spontaneously by buoyancy if a density inversion exists even from a perfectly flat source layer. But for solids to rise, some kind of imperfection is required to trigger growth, even though the source layer is potentially mobile.

This heterogeneity can be either in the source layer (the mother layer of buoyant material) or in its overburden. Lateral variations in thickness, density, viscosity, or temperature can all initiate upward movement of a buoyant source layer. Some geologic heterogeneities in the source layer are listed below.

Initial heterogeneities can be present in a salt layer as a result of accumulation in a graben in a rifted terrane or as a result of lateral facies changes (fig. 1). Heterogeneities may be imposed by faulting or folding of the source layer. Faulting can be basement involved or thin-skinned, as discussed in unit 10 (fig. 1). In each case the diapir is likely to form on the upthrown side of the fault where the load is less. Detached faults usually sole out within the ductile source layer, whether this is salt or shale.

In the case of fold-induced heterogeneities, the source layer also acts as the zone of detachment (fig. 1). Gravity gliding involves downslope sliding of strata and buckling of the frontal zone of the glide sheet. An example is the Mexican Ridges in the western Gulf, which are cored by shale. Gravity spreading involves slow, ductile collapse of rock under its own weight, and that of its overburden, and complementary lateral spreading. An example is the Sigsbee Scarp, where domes are initiated in the toe of a huge thrust nappe of salt spreading southward into the deep Gulf of Mexico. The best example of buckling from regional compression is the Zagros fold belt of Iran, formed by collision of the Arabian and Eurasian plates. The Precambrian Hormuz Salt provides a detachment zone for the folding cover. The folding (mid-Miocene to present) began long after the diapirs reached the surface and began to extrude salt (Jurassic to Late Cretaceous). Nevertheless, this folding is still acting like contracting bellows squeezing salt to the surface at an incredible calculated rate of 170 km/Ma (170 mm/year) at Kuh-e-Namak.

Key Reference: Woodbury and others (1980)

A. INITIAL INHOMOGENITIES

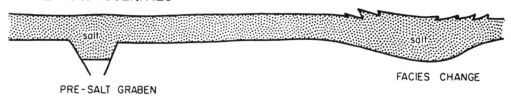

PRE-SALT GRABEN   FACIES CHANGE

B. IMPOSED INHOMOGENEITIES – FAULTING

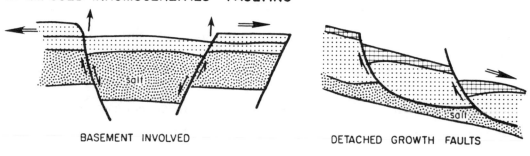

BASEMENT INVOLVED   DETACHED GROWTH FAULTS

C. IMPOSED INHOMOGENEITIES – FOLDING

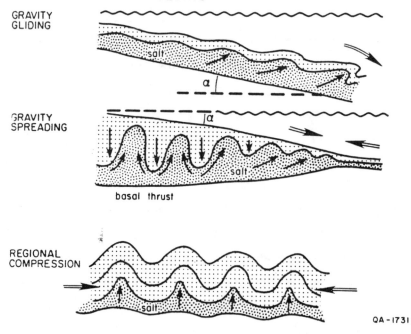

GRAVITY GLIDING

GRAVITY SPREADING

basal thrust

REGIONAL COMPRESSION

QA-1731

Figure 1. Initiation of upward-moving gravity structures at geologic inhomogeneities induced by lateral variations in thickness, density, viscosity, or temperature in buoyant source layer or cover. (Adapted from Woodbury and others, 1980.)

70

MECHANICS OF DIAPIR GROWTH

Flow Hydraulics in Idealized Systems

Overburden parallel to dipping source layer (fig. 1A). All layers are equally thick. The overburden pressure (lithostatic pressure) = λ (acceleration due to gravity) times the sum of all the products of layer thickness, η, and density, P, with correction for dip, a. The pressure head within the fluid source layer is equal at points P_1 and P_2 and everywhere else along the upper contact of the source layer. Thus the total hydraulic head between P_1 and P_2 equals only the difference in gravity head. Because P_1 is lower than P_2, salt flows downhill in the direction of lower total hydraulic head. In this example salt flows downhill regardless of the overburden density. Where both the source layer and its overburden are horizontal, the pressure head and gravity head are constant along a horizontal plane. The total hydraulic head is zero, so no flow occurs.

The geologic reaction to this flow is as follows. Because of the downdip flow from P_2 to P_1, the salt layer thins updip and thickens downdip, forming a wedge-shaped cross section. The updip pinch-out can even migrate downdip as a result of salt flow. A weak overburden also deforms because of the underlying flow; extension updip produces either ductile thinning or normal faulting, whereas shortening downdip produces ductile folding or brittle reverse faulting.

Inclined salt layer onlapped by sediments whose density increases downward (fig. 1B). The cover is terrigenous clastics compacting with increasing depth. The average density of units 1 and 2 is greater than that of the salt, whereas the average density of units 3 through 5 is less than that of salt, a realistic situation. The change in gravity head is constant along the top of the salt, so the tendency to flow downhill by gravity is constant. But the pressure head on top of the salt increases with depth because the thickness and density of the cover both increase toward the left. There is an increasing tendency to flow updip with greater burial, which counteracts the effect of the gravity head. The net result of the pressure head and gravity head is that the total head decreases downdip from P_3 to P_2, where the average density of the cover is exactly equal to the density of the salt. Downdip of P_2 the average density of the cover is greater than that at P_2. Because fluids flow from regions of high total head to low total head, salt will flow updip from P_1 to P_2, where the gravity head is exactly balanced by the pressure head.

The most important geologic result of this convergent flow to P_2 is the possible growth of a salt pillow there. However, considerable depth of burial would be required to initiate a pillow this way in a sequence of terrigenous clastic sediments, as we shall see in the following model.

Driving Upward Growth

Three principal models have been proposed as driving mechanisms for the rise of salt diapirs in noncompressive environments, a process known as halokinesis: buoyancy, differential loading, and thermal convection (fig. 2).

The Buoyancy model is the oldest and most widely known. The source layer in figure 2A is of low viscosity, but diapirs can also form in highly viscous rock like solid gneiss. However, the overburden must be of greater density than the source layer and it must be sufficiently yielding, as in the case of weakly consolidated sediments or even crystalline rocks buried in orogenic regions of high heat flow.

On burial the density of salt remains virtually constant, but clastic sediments grow denser by compaction and mineral phase changes (fig. 3). Under shallow burial, all terrigenous clastics are less dense than underlying salt--a positive density contrast because there is no density inversion. After about 600 m (2,000 ft) of burial, most sandy sequences exceed the density of the underlying salt; this constitutes a negative density contrast, or density inversion. Most shales compact to a density equivalent to salt after burial to 880 m (2,890 ft). Overpressured shales may not exceed the density of salt until burial to 1,800 m (6,000 ft) has compacted them sufficiently.

Once the density inversion exists, salt is potentially diapiric. Suppose that a thicker-than-average part of the salt layer exists, as shown with vertical exaggeration of 100 in figure 4. Above the dashed line, all densities and thicknesses are constant, so can be ignored. Below the dashed line, the differences involve the density of sediment enveloping (not overlying) the salt mound. In A, overburden such as carbonate is denser than salt. Pressure head at P_1 and P_2 is the product of overburden thickness (which is constant) and overburden density (which is not) times g. P_2 is overlain by denser rock than is P_1, so the pressure head is greater at P_2. Flow is toward P_1, and the salt pillow rises if the overburden is sufficiently yielding. In B an overburden like overpressured shale is less dense than salt. Because of the density differences, the pressure head is greater at P_1, so flow is toward P_3. This pillow would be evacuated of salt and would deflate.

This has important geologic implications. If evaporitic salt is succeeded directly by terrigenous sand or mud, any initial bulges in the salt surface tend to be squashed flat by the low-density cover because there is no density inversion. Conversely, if salt is succeeded by compact carbonates, or most chemical precipitates, any initial bulges of salt can be magnified by the high-density cover because of the density inversion.

Unlike buoyancy, the <u>Differential Loading model</u> does not require a density inversion and can operate when the salt is close to the surface. The most common agent of differential loading is the prograding delta (fig. 2B), which locally loads the underlying prodelta muds with great thicknesses of denser, coarser sediment. A prograding delta causes an upwelling of salt or shale around its margins, giving rise to a wide variety of isopach trends in overlying units, depending on the delta type. Other agents of differential loading are turbidite fans and depotroughs, sand barriers, oolite shoals, and lava flows. On the Texas-Louisiana continental slope, turbidite depotroughs fill 2,800-m (9,000-ft) deep synclines between rising salt diapirs. The turbidite load, which has compacted to a dense mass, causes the salt to flow from underneath the depotroughs into the diapirs, which causes further subsidence of the depotroughs. Although perpetuated and enlarged by differential loading and buoyancy, these particular diapirs were probably initiated by lateral squeezing of the Sigsbee salt mass or by thermal convection at its base.

<u>Relative effectiveness of buoyancy and differential loading</u>: If salt is buried beneath a denser cover, how do the mechanisms of buoyancy and differential loading compare in effectiveness in driving diapirism? Assume a uniformly layered sequence containing a mound 20 m high and 10 km wide (or 100 ft high and 10 mi wide) (fig. 5A). In A the mound is a salt pillow enveloped by denser material with density equal to the average Gulf Coast sediment at 1,200 m (4,000 ft) depth (2.25 g/cm^3). By calculating the total hydraulic head at P_1 and P_2, we can convert this to a flow gradient of $2 \times 10^{-4} \lambda$ (gravity acceleration), based on the slope of the pillow flanks. With an extreme density inversion where salt underlies a dense, mixed anhydrite-carbonate sequence (density 2.75 g/cm^3), the flow gradient is increased to $2 \times 10^{-3} \lambda$.

In figure 5B the mound is a delta prograding toward the viewer and loading the sediments and salt beneath it. If the average density of the delta is 2.0 g/cm^3 and that of seawater is 1.0 g/cm^3, the flow gradient is $4 \times 10^{-3} \lambda$. This gradient is 20 times that induced by buoyancy of salt beneath sediments of average density, and twice that induced by buoyancy beneath the densest overburden (anhydrite-carbonate mixture). As a driving mechanism for dome growth, therefore, differential loading is much more effective than buoyancy. This is because for the same size heterogeneity, the density contrast between seawater and deltaic sediments (2.00) is much larger than that between salt and its enveloping overburden of terrigenous clastics (1.02).

Thermal Convection is the third model for halokinesis. Unlike buoyancy and differential loading, thermal convection can operate with no overburden to the salt layer. With increasing depth, the salt becomes hotter and weaker and expands due to the heat. This expansion is greater than the contraction due to increasing pressure; at 5,000 m (17,000 ft) depth under an average geotherm of 30°C/km (1.7°F/100 ft) halite expands 2 percent due to heat and contracts 0.5 percent due to pressure. Salt, therefore, becomes less dense with burial, in contrast to all other common sediments. Under this geotherm, the Rayleigh Number exceeds a critical value (around 650) in salt layers more than 1,500 m (5,000 ft) thick. Thus low-density, hot solid salt rises by thermal convection from the base of the salt layer. Salt rising from a depth of 1,500 m (5,000 ft) has a temperature of about 70°C (160°F), well below the melting point of 800°C (1,475°F). If fresh meteoric water is added, it reduces the yield strength and accelerates rise of the salt. In Ethiopia's Danakil depression, volcano-shaped mounds of salt up to 60 m (200 ft) high emit hot springs that may be the tops of convecting plumes of salt. Convection does not require an overburden but will nevertheless be enhanced by a sedimentary cover which acts as an insulating blanket, allowing heat to build up in the salt beneath it. In a 1,000-m (3,000-ft) thick salt layer beneath 3,000 m (10,000 ft) of overburden, thermal convection is theoretically possible. Under the present heat flow in the Gulf Coast, hot (but not molten) salt could still be rising in deep salt pillows and in the base of diapirs.

In addition to the three driving mechanisms described above, salt can be passively deformed by strain in a strong overburden due to stress unconnected with salt flow. For example, salt can flow passively against gravity into low-pressure zones in the cores of anticlines formed by regional compression (fig. 2D).

Different Mechanisms Drive Salt Under Different Conditions

Shallow Depths (<600 m, <2,000 ft):

1. Differential loading is the main agent of dome growth, especially during delta progradation.

2. Buoyancy can operate if the cover is initially dense, e.g., compact carbonates.

3. Tectokinesis. Passive rise of salt into anticlines deformed by gravity-glide folding is possible in the lower parts of the continental slope or marginal basin.

4. Thermal convection is possible where salt is thick (>1,500 m, >5,000 ft) and heat flow is high (e.g., 30°C/km).

Deeper than 600 m, 2,000 ft:

1. Terrigenous clastic overburden has compacted so that its density exceeds that of salt, so buoyancy is increasingly effective beneath any overburden.

2. Differential loading in salt-withdrawal areas continues to drive diapirism.

3. Gravity spreading of a thick salt mass basinward beneath a prograding continental margin encourages diapirism on the continental slope.

4. Thermal convection probably accelerates salt diapirism and is increasingly effective with depth but decreasingly effective if the crust is cooling.

Key References: Kehle (in preparation), Talbot (1978)

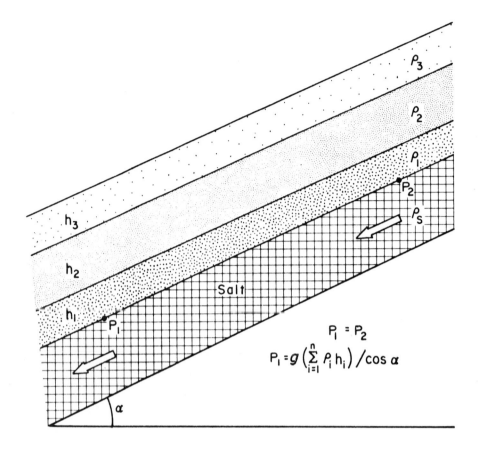

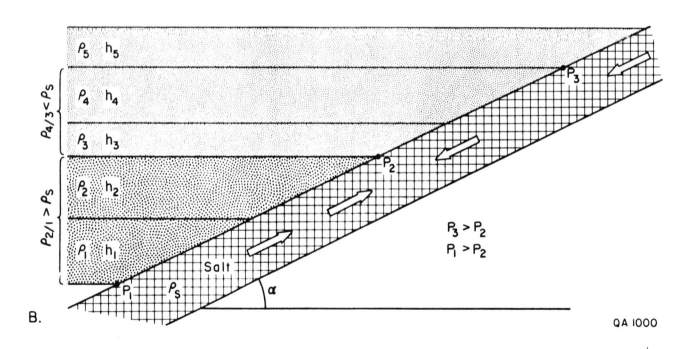

Figure 1. Flow hydraulics in a tilted source layer of salt. A. Tabular overburden parallel to source layer. B. Wedgelike overburden onlapping source layer. P_1, P_2, and P_3 are points in the system and pressures at those points; h_1 through h_5 are stratigraphic thicknesses of overburden units; ρ_s and ρ_1 through ρ_5 are mean densities of the source layer and overburden units; α is the dip of the source layer. (Modified from R. O. Kehle, unpublished data.)

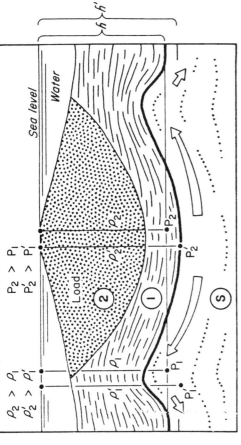

A. BUOYANCY HALOKINESIS

$\rho_3 > \rho_2 > \rho_1$
$P_3 > P_2 > P_1$

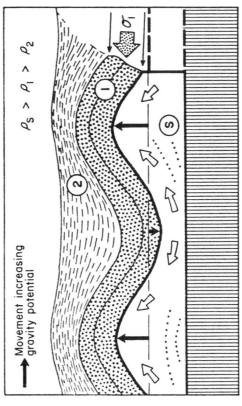

B. DIFFERENTIAL-LOADING HALOKINESIS

$\rho_2 > \rho_1$
$\rho_2' > \rho_1'$

$P_2 > P_1$
$P_2' > P_1'$

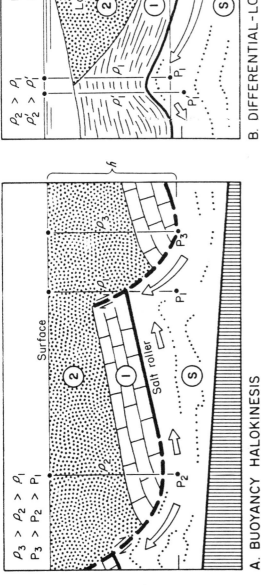

C. THERMAL CONVECTION HALOKINESIS

$T_2 > T_1$
$\rho_2 < \rho_1$

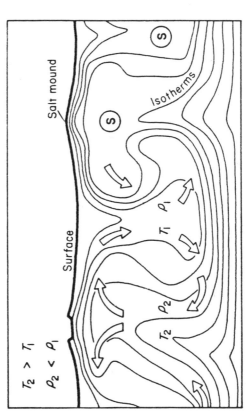

D. TECTOKINESIS

$\rho_s > \rho_1 > \rho_2$

Figure 2. Four principal mechanisms for salt flow; they may combine in nature. Symbols as in figure 1. A. Buoyancy halokinesis: the formation of a salt roller by normal faulting (extensional tectokinesis) provides the initial geological inhomogeneity required for buoyant salt to rise above P_1 because the mean density above there is minimal. B. Differential loading halokinesis: salt rises above P_1 because salt at P_2 is driven sideways beneath the extra thick load of dense sediment. C. Thermal convection halokinesis: geothermal gradient causes the base of a thick layer of salt to heat sufficiently for thermal convection to occur; hot (temperature T_1) expanded salt rises in plumes, perhaps with surface expression as mounds, and sinks at the edges of convection cells after cooling (temperature T_2). Tectokinesis: salt rises passively against the force of gravity in anticlines and is displaced beneath synclines during detachment folding of light cover under regional compressive stress; if a density inversion were present, buoyancy halokinesis would augment the tectokinesis, as in A.

77

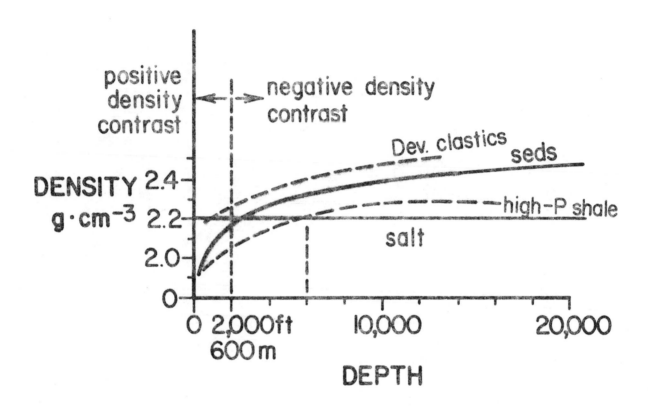

Figure 3. Relation between burial depth and density in salt and terrigenous clastic sediments.
(Adapted from Nettleton, 1934; Dickinson, 1953; Gussow, 1968.)

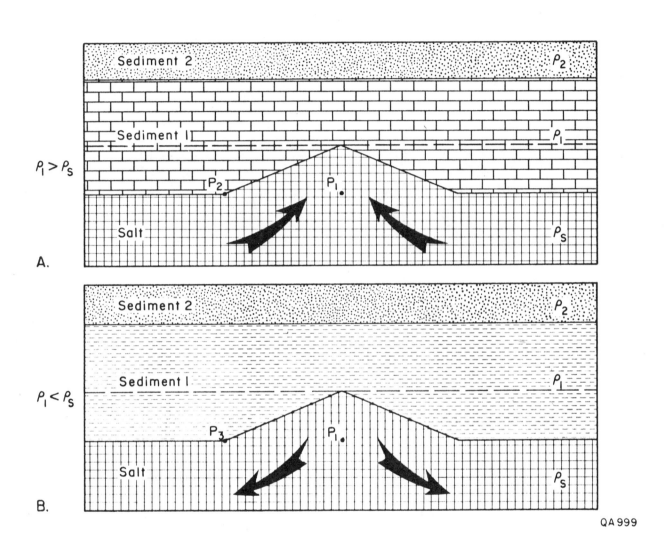

Figure 4. A. Rise of a salt mound beneath denser cover. B. Collapse of a salt mound beneath less dense cover. (Adapted from R. O. Kehle, unpublished data.)

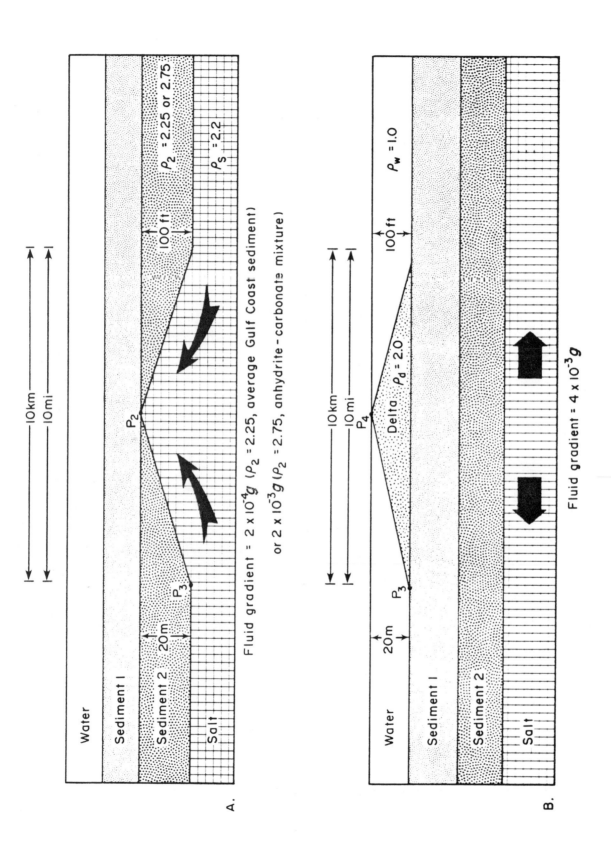

Figure 5. Comparison of the effectiveness of buoyancy (A) versus differential loading (B) as a halokinetic agent. For the same size of structure, differential loading causes a higher flow gradient and is therefore more effective than is buoyancy. (Adapted from R. O. Kehle, unpublished data.)

80

DOME-GROWTH STAGES

As a salt structure evolves, it passes through three stages of growth, known as the pillow stage, the diapir stage, and the postdiapir stage. The model shown in figures 1 and 2 assumes that all thickness variations in strata are primary and caused by syndepositional rise or fall of the sedimentation surface because of salt flow beneath. Evidence suggests that once initiated, the crest of a salt diapir stays more or less at a constant shallow depth while its base and the surrounding strata sink as the basin floor subsides, a process known as downbuilding.

When salt evolves from a tabular layer to a diapir, three things change.

1. Salt Shape

The salt mass changes from planar to pillow (salt mound with fairly planar base, overlying strata concordant) to diapir (salt stock piercing overburden strata).

2. Structural Inversion

The diapir originally rises from a salt pillow or ridge overlain by strata forming a regional anticline. Because of deflation of the pillow during diapir growth, the regional anticline collapses and inverts to a regional syncline containing a local protuberance--the salt dome itself. Adjacent to the growing salt structure, a regional syncline inverts to an anticline when its ends that are nearest two salt domes subside as the salt pillow deflates to feed the diapir. The resulting structure--a turtle-structure anticline--has a planar base and domal crest (fig. 3). It can be underlain by salt, but it owes its thickness to clastic sediments that originally accumulated in the peripheral sink during pillow growth.

3. Thickness Changes

If the salt remains quiescent during burial, overlying strata are of uniform thickness on a regional scale (disregarding local lapouts and thickness changes caused by basin-wide processes) (fig. 1A). The pillow stage is recognized by thinning of strata toward the dome; thick sediments accumulate in the primary peripheral sink (fig. 1B, 2A). The term "peripheral sink" is a genetic one, referring to a feature formed in a specific way. The term "rim syncline" is a purely structural one for a closed (roughly circular) syncline around a dome. In figure 2A the rim syncline coincides with the primary peripheral sink, but not in figure 2B or 2C.

The diapir stage of growth is recognized by thickening of strata toward the dome into the secondary peripheral sink (figs. 1D and 2B). This sink forms as the underlying salt pillow inverts to become a turtle-structure anticline.

The onset of the postdiapir stage marks the significant point where the diapir exhausts its underlying reservoir of salt. If the diapir continues to rise relative to the surrounding strata, it must elongate by thinning of its lower trunk, or even by complete detachment like an inverted teardrop. This can be recognized by two features, not necessarily occurring at the same time.

1. Assuming the regional dip is zero for simplicity (corrections can be made for significant regional dip), the level at which surrounding strata have zero dip (outside the narrow aureole of upward drag adjacent to the salt stock) marks the point where the rate of salt withdrawal in the rim syncline has declined to zero. This level commonly coincides with the level of maximum overhang if overhang is present. Thus if drilling some distance from the salt stock encounters a steepening of strata just below strata with regional dip, this is the most favorable level for an overhang to form and is an attractive area in which to deflect the drillhole toward the dome to test for an overhang trap.

2. The level above which sedimentary units in the tertiary peripheral sink do not thicken toward the diapir also indicates cessation of salt withdrawal from beneath the sink.

Effects of Viscosity Contrast

Experimental and mathematical modeling has shown that ultimate dome shape and its capacity to trap hydrocarbons are strongly controlled by viscosity contrasts between the source layer (e.g., salt) and its cover (surrounding and overlying sediments). Let M be the ratio between the viscosity of the cover and the viscosity of the buoyant source layer (fig. 4). Salt domes are usually drawn as in A, which implies that the viscosity of salt is much higher than that of its clastic cover. But the consensus is that salt deforms as a softer substance than the cover, for the present viscosity of Gulf Coast overburden is estimated at 10^{21}P, whereas that for salt varies from 10^{14} to 10^{17}P (depending on dryness). If these viscosity values are correct, we have to wonder whether salt domes of the Gulf Coast do not have shapes and internal circulation patterns more like those in C. These huge overhangs would be effective traps, but this part of a salt stock is poorly known at present because it lies inside the zone normally drilled. Conventional wisdom dictates that a steep wall of salt will be found a little way beneath the overhang, whereas the model in C suggests that reservoirs might be found much closer to the axis of the dome within the seismic reflection shadow. Perhaps we should look at our existing gravity and drill data to see whether diapirs have much narrower stems than we think. Only structurally mature domes have these different shapes; immature domes are of similar shape, but are more closely spaced where M is high.

Effects of Inclination on Overhang

Experiments have also shown that where a source layer is inclined, as it is on the edges of a large basin like the Gulf of Mexico, diapirs may preferentially overhang the downdip side (fig. 5). Other factors may also influence the direction of maximum overhang, but statistically we can expect more overhangs on the basinward side of diapirs if they grew from an inclined source layer, a useful guide to well siting. The diapir initially grows symmetrically normal to its source layer, rather than vertically, as might be expected (fig. 6). An example of preferential overhang of natural diapirs is shown in figure 7.

Overhangs form because the top of the stock reaches:

1. A free surface below water (in which it dissolves) or air (where it might extrude as a salt glacier or might dissolve depending on rainfall).
2. An impenetrable stratum beneath which it spreads.
3. A layer with equal or lower density within which it spreads.
4. A layer of lower confining pressure within which it spreads.

Key References: Seni and Jackson (1983a, b), Trusheim (1960)

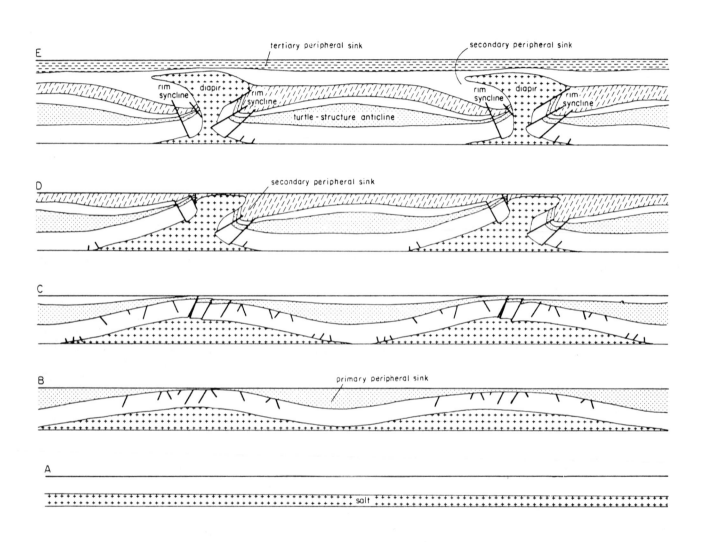

Figure 1. Structural evolution of salt pillows, salt diapirs, their peripheral sinks, and a turtle structure. (From Seni and Jackson, 1983a, adapted from Trusheim, 1960.)

GROWTH STAGE	UPLIFTED AREA	WITHDRAWAL BASIN

PILLOW

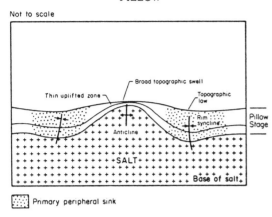

Primary peripheral sink

UPLIFTED AREA

Geometry

Sediments above pillow are thin over broad, equidimensional to elongate area. Maximum thinning over crest. Area extends 100 to 400 km² (40 to 150 mi²), depending on size of pillow. Percentage thinning, 10 to 100%.

Facies

Thin, sand-poor, fluvial-deltaic deposits over crest of pillow include interchannel and interdeltaic facies. Erosion common. Carbonate deposits on crest would include reef, reef-associated, and high-energy facies.

WITHDRAWAL BASIN

Geometry

Sediments are overthickened in broad to elongate primary peripheral sink, generally located on updip side of salt pillow. Axial trace of sink parallels axial trace of elongate uplift, generally separated by 10 to 20 km (6 to 12 mi). Sink attains 300 km² (120 mi²) in extent, depending on size of pillow. Percentage thickening, 10 to 30%. Recognition of primary peripheral sink may be hindered by interference of nearby salt structures.

Facies

Thick, sand-rich, fluvial-deltaic deposits in primary peripheral sink include channel axes and deltaic depocenters. Aggradation common in topographically low area of sink. Carbonate deposits in sink would include low-energy facies caused by increase in water depth.

DIAPIR

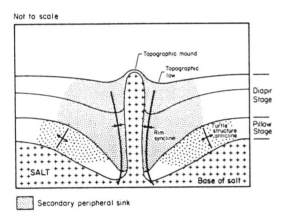

Secondary peripheral sink

UPLIFTED AREA

Geometry

Strata largely absent above dome. An 8 to 50 km² (3 to 20 mi²) area around diapir is thinned, depending on size and dip on flanks of dome.

Facies

Facies immediately over dome crest not preserved because of piercing by diapir of all but the youngest strata. Sand bodies commonly pinch out against dome flanks.

WITHDRAWAL BASIN

Geometry

Sediments are thickened up to 215% in secondary peripheral sink. Sinks up to 1,000 km² (390 mi²) in extent are equidimensional to elongate, and they preferentially surround single or multiple domes; several sinks flank domes; percentage thickening ranges from 50 to 215%.

Facies

Expanded section of marine facies dominates, including limestones, chalks, and mudstones; generally sink is filled with deeper water low-energy facies caused by increased water depth. Elevated saddles between withdrawal basins are favored sites of reef growth and accumulated high-energy carbonate deposits.

POST-DIAPIR

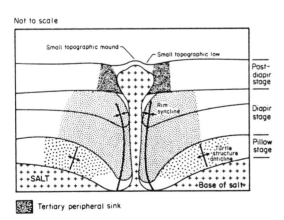

Tertiary peripheral sink

UPLIFTED AREA

Geometry

Strata thin or absent in small 10 to 50 km² (4 to 20 mi²) area over crest and adjacent to dome; area depends on size of dome and dip of flanks.

Facies

Facies and strata over crest of dome not preserved in places of complete piercement. Modern analogs have interchannel and interdeltaic facies in uplifted area. Mounds above dome include thin sands. Carbonate strata would include reef or high-energy deposits; erosion common.

WITHDRAWAL BASIN

Geometry

Sediments within 20 to 200 km² (8 to 80 mi²) tertiary peripheral sink are thickened 0 to 40%, commonly by < 30 m (100 ft). Axial trace of elongate to equidimensional sink surrounds or flanks a single dome, or connects a series of domes.

Facies

Modern analogs have channel axes in sink. Aggradation of thick sands common in subsiding sink. Carbonate strata would include low-energy facies.

Figure 2. Stages of salt-dome growth showing typical lithologic and thickness variations in strata around domes. (From Seni and Jackson, 1983a.)

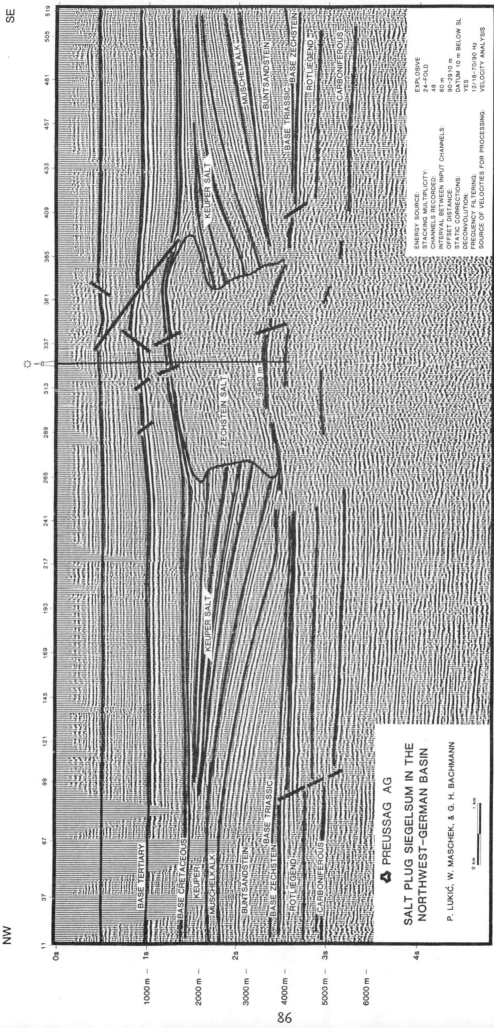

Figure 3. Two turtle structures cored by the Buntsandstein flank Siege.sum Dome, northwest Germany. Diapirism began in Muschelkalk time, initially withdrawing salt from the NW and subsequently withdrawing salt from the SE in Keuper time. The presence of gas indicates an anticline below the salt stock, notwithstanding possible velocity pullup of the seismic reflections. (From Lukić and others in Bally, 1983.)

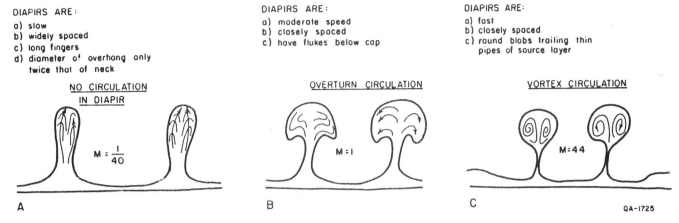

DIAPIRS ARE:
a) slow
b) widely spaced
c) long fingers
d) diameter of overhang only
   twice that of neck

NO CIRCULATION
IN DIAPIR

$M = \frac{1}{40}$

A

DIAPIRS ARE:
a) moderate speed
b) closely spaced
c) have flukes below cap

OVERTURN CIRCULATION

M = 1

B

DIAPIRS ARE:
a) fast
b) closely spaced
c) round blobs trailing thin
   pipes of source layer

VORTEX CIRCULATION

M = 44

C

QA-1725

Figure 4. Effects of viscosity contrast on model dome shape and internal circulation patterns based on experimental and numerical modeling by Berner and others (1972), Whitehead and Luther (1975), Woidt (1978), and Heye (1978, 1979), and unpublished work at the University of Uppsala. M is the ratio of overburden viscosity to dome viscosity. A. Slow rising, widely spaced, thumb-shaped diapirs with cap diameter only up to twice that of stem; no internal circulation of diapiric material. B. Moderately fast rising, closely spaced, mushroom-shaped diapirs with broad caps having peripheral hanging lobes filled by downward-facing salt tongues. C. Fast rising, closely spaced diapirs comprising round blobs with internal vortex circulation trailing thin pipes of source layer; peripheral sinks form.

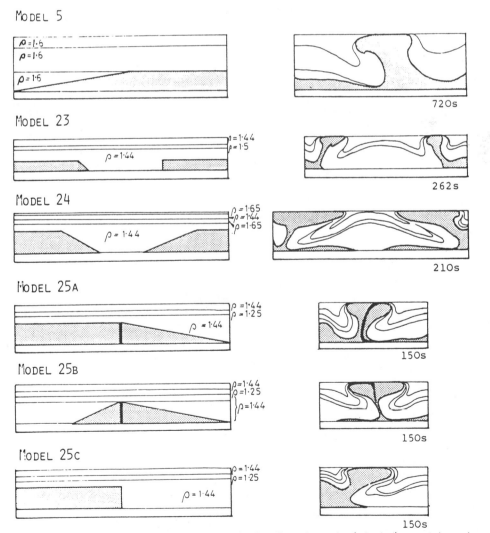

MODEL 5

$\rho = 1.6$
$\rho = 1.6$
$\rho = 1.6$

720s

MODEL 23

$\rho = 1.44$
$\rho = 1.44$
$\rho = 1.5$

262s

MODEL 24

$\rho = 1.44$
$\rho = 1.65$
$\rho = 1.44$
$\rho = 1.65$

210s

MODEL 25A

$\rho = 1.44$
$\rho = 1.44$
$\rho = 1.25$

150s

MODEL 25B

$\rho = 1.44$
$\rho = 1.44$
$\rho = 1.25$
$\rho = 1.44$

150s

MODEL 25C

$\rho = 1.44$
$\rho = 1.44$
$\rho = 1.25$

150s

Figure 5. Inclined and asymmetric model domes forming where the (stippled) source layer has thickness discontinuities; note the preferred overhang on the downdip side of inclined source layers. (From Talbot, 1977.)

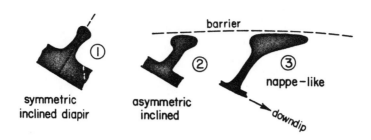

Figure 6. Summary of the growth of inclined model domes, based on figure 5 and related experiments.

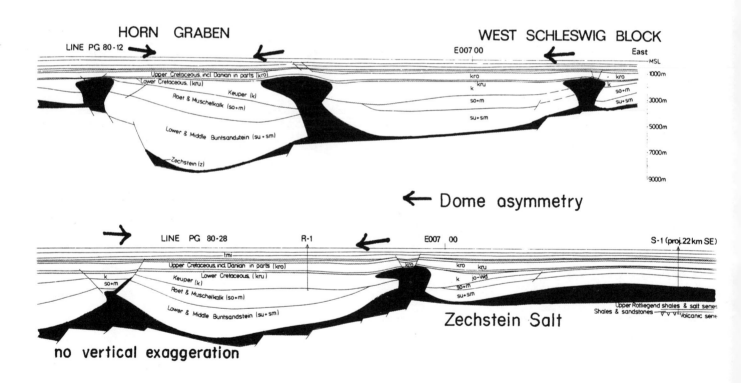

Figure 7. Zechstein salt diapirs in the Horn Graben, southeastern North Sea, showing preferred overhang toward the downdip side, as predicted by model experiments. Angular unconformities at the base of the Keuper around the central dome in each line indicate widespread erosion or non-deposition during the final stages of pillow uplift just before diapirism began. (Adapted from Best and others, 1983.)

CONTOUR INTERACTIONS BETWEEN REGIONAL DIP, SALT WITHDRAWAL, AND UPWARD DRAG

Every surface pierced by a salt dome has a unique structure, so that every structure-contour pattern around a dome is different. Nevertheless, the structure around domes is governed by the same set of factors--factors that we must be familiar with if we are to create geologically realistic structure maps from a minimum of well data. Fault patterns are discussed in unit 19. Thus, although faulting is an important component in structure-contour maps, there is no need for repetition here. Let us examine the folding associated with a salt dome in terms of the structure-contour pattern.

Folding around Domes

Figure 1 shows the cross section and contour pattern of a surface that shows the separate effects of a 1.3° regional dip, 300 m (1,000 ft) of symmetric salt withdrawal, and 300 m (1,000 ft) of symmetric, upward drag.

Regional dip characterizes virtually all continental margins and the flanks of basins in the continental interior. Its contour pattern requires no introduction.

Salt withdrawal results from the deflation of a salt pillow as salt moves from it up a diapir toward the surface, where it can be attacked by dissolution in freshwater aquifers or even by erosion if it becomes exposed. Salt withdrawal from a pillow causes a salt-withdrawal basin around the salt stock. The deepest part of the withdrawal basin (peripheral sink) is closest to the salt stock, where the rate of salt withdrawal is greatest. A concentric target pattern results, the innermost contour being the lowest.

Upward drag of strata adjacent to the salt stock results from friction between the rising salt and the sinking sediments around it. Stratal dip increases exponentially toward the salt, and the highest part of the upward drag zone is at the salt contact. A concentric target pattern results, the innermost contour being the highest.

What concerns us is how regional dip, salt withdrawal, and upward drag interact structurally, for all these elements are present in the real world.

The combination of regional dip and salt withdrawal produces a peripheral sink that is semicircular or crescent-shaped (fig. 2). The axial trace of the rim syncline passes through the deepest part of the sink and completely encircles the salt stock, even though its presence is subtle on the side of the dome opposite the sink. A monoclinal rim anticline marks the outer limit of dome-specific deformation.

The combination of regional dip, salt withdrawal, and upward drag forms a crescentic peripheral sink, through which the trace of the rim syncline runs, encircling the dome (fig. 3). A rim anticline, which marks the edge of the withdrawal basin, encircles the rim syncline. But dome-associated deformation extends beyond the anticline because of the broad aureole of upward drag.

Deducing the Direction of Maximum Salt Withdrawal

The most important feature of the structural patterns in figures 2 and 3 is the position of the crescentic withdrawal basin. It lies on the downdip side of the diapir and its deepest part is directly downdip.

If the deepest part of the peripheral sink (trough point) is not situated directly down regional dip of the underlying diapir, salt flow must have been asymmetric, roughly from the direction in which the trough point lies. Salt is commonly preferentially drawn from the updip side of the diapir, probably because the regional flow of salt in the source layer was downdip (refer to section on Mechanics of Salt Flow, unit 16). Knowing this asymmetry, we can deduce that this is the most probable direction in which a large (in terms of structural closure or relief) turtle structure lies at a deep level.

Implications to Petroleum Migration

Theoretically, if the folded unit defining the structure contains hydrocarbons before it is deformed (probably unlikely in view of the syndepositional nature of dome growth and sedimentation), hydrocarbons around the dome begin to migrate locally in response to folding (fig. 4). Oil and gas inside the axial trace of the rim syncline migrate inward to the flanks of the diapir; oil and gas outside the axial trace migrate updip toward the outer rim anticline. Here they accumulate if another withdrawal basin is present to reverse the regional dip, as shown in figure 4.

If petroleum migrates up the regional dip toward the dome <u>after</u> the formation of the folds in figure 4, on reaching the outer rim anticline the fluids tend to be deflected from the diapir. No hydrocarbons are trapped by the dome. Instead, oil and gas accumulate, as in the previous scenario, in the residual high between two adjacent withdrawal basins (fig. 4).

However, in the Gulf Coast abundant hydrocarbons have been trapped close to salt diapirs. This suggests that either petroleum migrated very early before the deflecting structures had grown significantly, or that we are neglecting the third dimension. If the latter is true, lateral migration is subordinate to vertical migration around domes, suggesting that hydrocarbons migrate up the flanks of the dome vertically, perhaps exploiting the fractured aureole around a salt stock, until they are trapped at higher levels.

<u>Key Reference</u>: Ritz (1936)

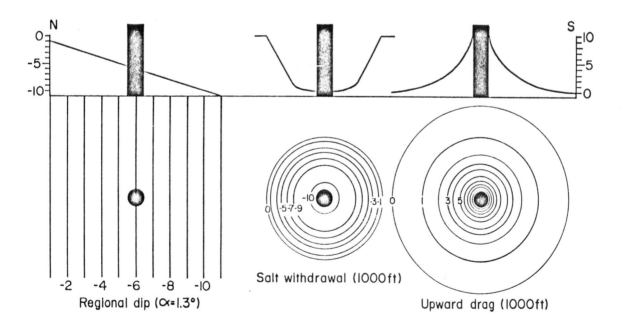

Figure 1. Cross sections (above) and structure-contour maps (below) of an arbitrary horizon around a salt diapir (solid), showing the separate effects of regional dip, symmetrical salt withdrawal, and symmetrical upward drag.

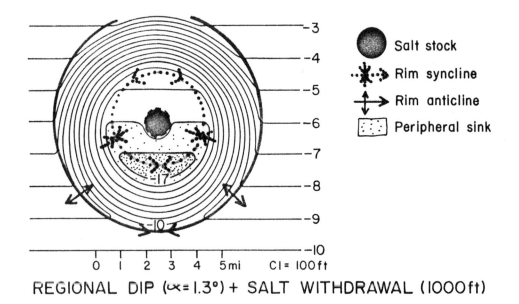

REGIONAL DIP (∝=1.3°) + SALT WITHDRAWAL (1000ft)

Figure 2. Structure-contour map showing the combined effects of regional dip and salt withdrawal in figure 1. (Adapted from Ritz, 1936.)

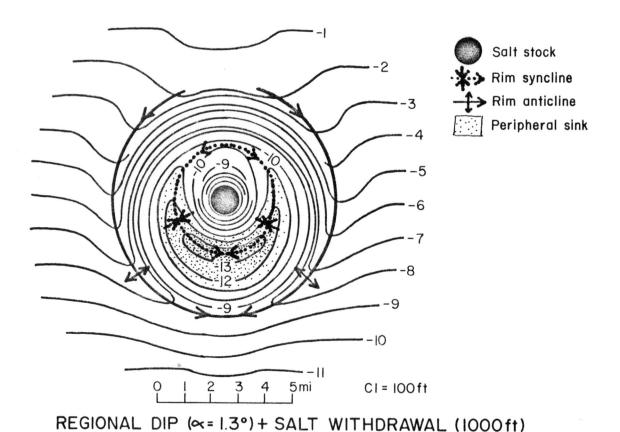

REGIONAL DIP (∝=1.3°) + SALT WITHDRAWAL (1000ft)
+ UPWARD DRAG (1200ft)

Figure 3. Structure-contour map showing the combined effects of regional dip, salt withdrawal, and upward drag in figure 1. (Adapted from Ritz, 1936.)

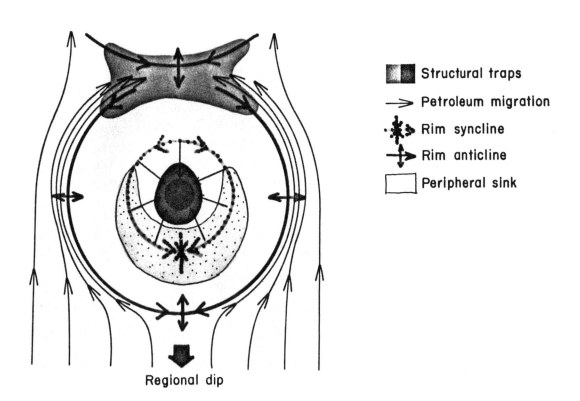

Structural traps

⟶ Petroleum migration

Rim syncline

Rim anticline

Peripheral sink

Regional dip

Figure 4. Petroleum migration and structural traps in the structure shown in figure 3. (Adapted from Ritz, 1936.)

FAULT PATTERNS AROUND SALT DOMES

The complex faulting above and around salt stocks creates many small reservoirs, the exploitation of which requires good well control and good understanding of the principles of faulting around domes. The two most important factors that can influence fault patterns near surface are the plan shape of the salt stock and the type of regional strain. The influence of both these factors has been investigated in plan view by experimental and mathematical modeling of the effects of gentle doming.

Faults in Plan View

Without regional strain during doming, circular domes develop normal faults radiating from the center of the dome. These faults are most abundant and their orientation is most radial on the dome flanks, rather than on the crest (fig. 1). Theoretically, peripheral concentric zones of strike-slip faults and reverse faults are possible, but were not observed in the experiments (fig. 2A). Elliptical domes initially develop normal faults on their crests, roughly parallel to their long axes, but splaying outward at the narrow ends of the dome where curvature is greatest (figs. 1 and 3A). With further uplift, radial normal faults form farther from the crest line of the elliptical dome. Theoretically zones of strike-slip and reverse faulting are possible, but these were not observed in the experiments (fig. 3A).

With regional extension during doming, typical of the Gulf Coast environment, normal faults form on the crest and flanks of circular domes; some faults have a component of strike-slip near the margin of the domes. With regional extension at the same rate as uplift, normal faults form perpendicular to the axis of maximum extension in the crest, where both horizontal stresses are tensile (fig. 2B). With regional extension twice as fast as uplift, almost all the faults are parallel to each other and perpendicular to the direction of maximum extension. Strike-slip faults trending 60° from the regional extension direction can form near the periphery of the dome (figs. 1 and 2BC). In the case of elliptical domes, the fault pattern is similar to that of elliptical doming without regional extension, except for small differences such as greater parallelism of crestal normal faults with the dome long axis and, of course, many parallel normal faults beyond the domed area. If the long axis of the dome is parallel to the regional extension direction, normal faults form perpendicular to the dome long axis (fig. 4).

With regional shortening during doming, normal faults form on the crest of circular domes, mostly parallel to the direction of maximum regional shortening. Strike-slip and reverse faults form near the edges of the domes about 30° and 90°, respectively, from the shortening direction. Under regional shortening elliptical domes form patterns of faults quite different from the other states of regional strain. Normal faults trend parallel to the maximum shortening direction, or normal to the long axis of the dome, a parallelism that increases as the intensity of shortening increases. With increasing uplift, reverse faults and strike-slip faults form near the edge of the elliptical dome.

It should be emphasized that the above analyses apply only to near-surface faults above gentle domes in mechanically homogeneous material. With a greater intensity of doming, such as immediately above a diapir, we can expect ring faults to form tangentially to the stock to accommodate the strong differential uplift of the stock.

Faults in Cross Section

In order to analyze the type and orientation of faults above salt domes in cross section, we have to consider the geometry of conjugate normal faults. For each stress regime a conjugate pair of faults is possible. Conjugate faults form symmetrically about the maximum compressive stress, which bisects their acute dihedral angle. They intersect along the intermediate stress axis. Because both faults in the set require the same stress orientation, we would expect them to operate simultaneously or alternately.

To realistically model this simultaneous or alternate movement along faults above a dome is complex. Many geologists take the easy way by making one orientation of fault consistently offset another, even if this contradicts stratigraphic relations (fig. 5). Or they do not even consider the problem, merely deciding by whim which fault offsets another, possibly creating long throughgoing faults that have no basis in reality.

Experimental modeling shows that crossing conjugate faults can operate simultaneously (fig. 6). Where they cross, neither fault offsets the other if movement is negligible. With further displacement, both faults are mutually offset. Once a fault is offset, it becomes locked, and further extension forms new conjugate faults that may be steeper than, shallower than, or parallel to their locked predecessors (fig. 7). Generally in a crestal graben bounded by sets of conjugate faults, younger sets form closer to the axis of the graben.

Key References: Horsfield (1980), Withjack and Scheiner (1982)

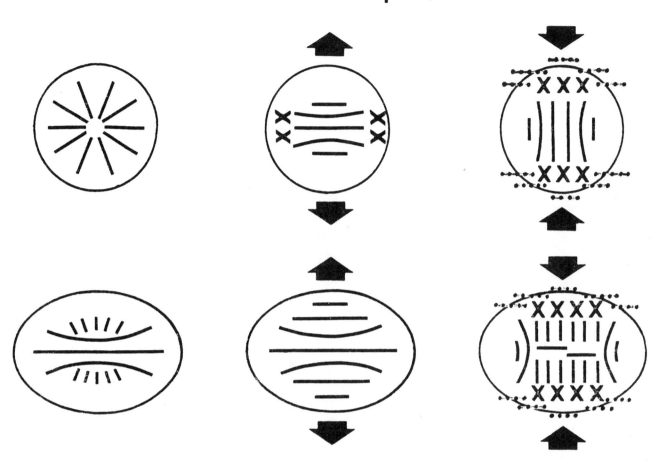

— Normal ➤Strike-Slip ••••Reverse

Figure 1. Fault patterns produced in experimentally modeled domes. Arrows indicate superimposed regional stresses during doming. (Adapted from Withjack and Scheiner, 1982.)

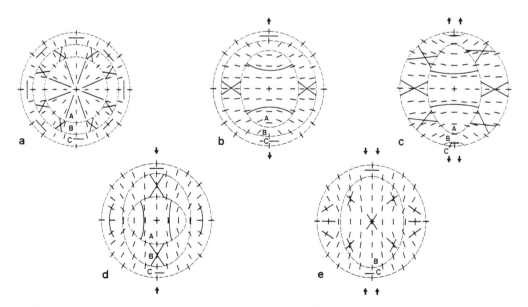

Figure 2. Calculated surface fault patterns (thick, solid lines) for circular domes. Zone A, normal faults; zone B, strike-slip faults; zone C, thrust faults. Single arrows indicate regional extension or shortening at same rate as doming; double arrows indicate regional strain at twice the rate of doming. (From Withjack and Scheiner, 1982.)

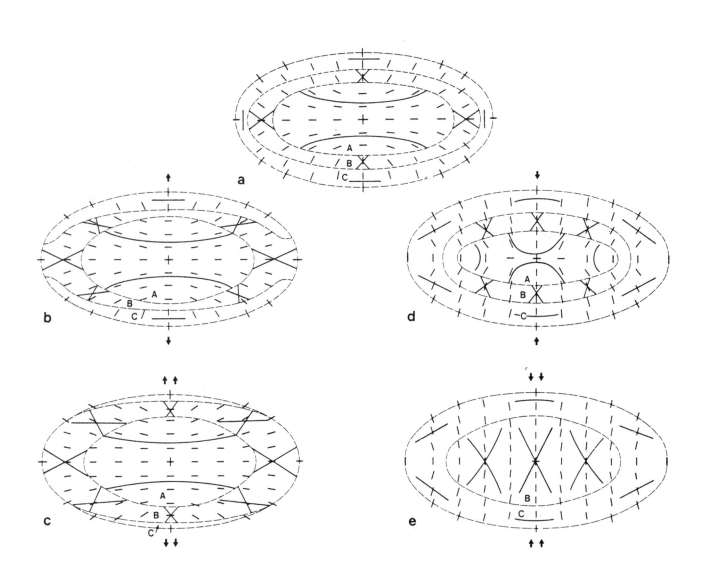

Figure 3. Calculated surface fault patterns of elliptical domes. (From Withjack and Scheiner, 1982.)

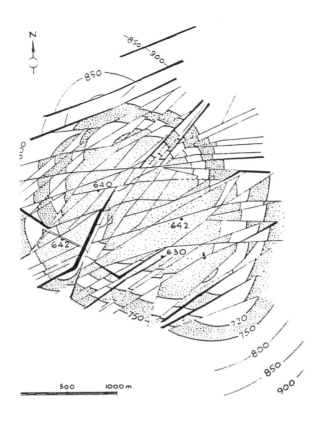

Figure 4. Subsurface normal faults showing strong NE alignment over Rietbrook Dome, North Germany. Regional extension was NW-SE. (From Cloos, 1968, after Behrmann, 1949.)

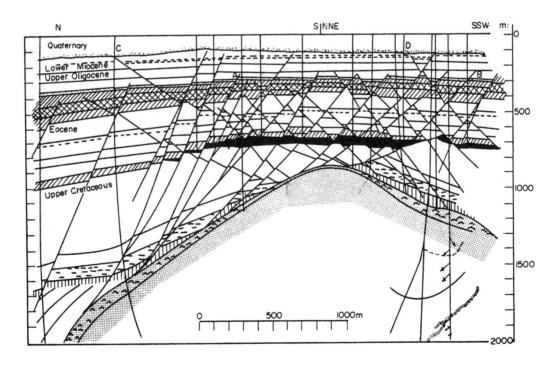

Figure 5. Cross section through Rietbrook Dome, North Germany, showing one interpretation of the fault geometry. (From Horsfield, 1980, after Behrmann, 1949.)

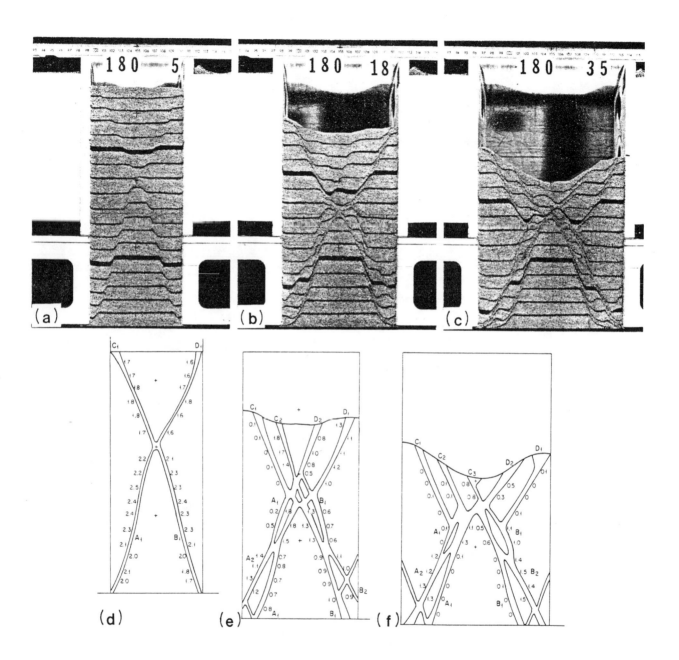

Figure 6. Sandbox experiment of contemporaneous movement along crossing conjugate faults in photographs (above) and line drawings (below) showing values of measured throw in mm after lateral extension of 5 mm, 18 mm, and 35 mm. (From Horsfield, 1980.)

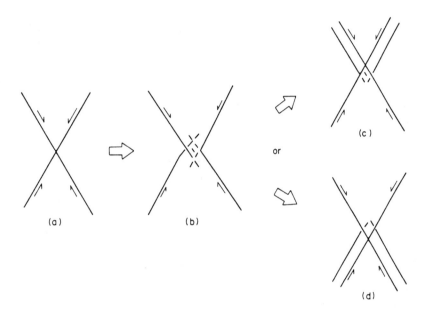

Figure 7. The proposed mechanism whereby a pair of crossing conjugate faults can move contemporaneously. A. Initial fracture. B. Mutual offset as faults move together. C. and D. Alternative patterns produced by locking of either the upper or lower halves of each conjugate pair and continued activation of the other half. (From Horsfield, 1980.)

PETROLEUM TRAPS AND ASSOCIATED FACIES OF SALT DOMES

It hardly needs to be restated that oil fields associated with salt diapirs have constituted the most important play in the Gulf Coast area since the discovery of Spindletop in 1901. In the petroleum industry the term "piercement dome" is restricted to salt diapirs that have pierced a reservoir section. Under this restricted usage, many deep diapirs are not termed piercement although they are true diapirs apart from the lack of reservoirs around them. This restricted usage is not always adhered to, as in the depth classification of "piercement domes" in figure 1; in relation to the reservoir, the deep dome shown is nonpiercing.

The commonest types of the many traps associated with salt domes are now well known (fig. 2). Most of these traps are structural. Here we describe some of the largely stratigraphic subtle traps related to salt domes.

In dealing with the growth stages of salt domes, we saw how each of the three stages of growth is characterized by particular thickness changes and facies types in the surrounding sediments that accumulated during that stage of growth. The geometry of some of these facies enables them to act as reservoirs. Even though some of these facies can be far from the actual salt dome, they owe their origin to the influence of the growing dome on sedimentation patterns, and are therefore dome-related.

Facies developed during a particular stage of dome growth depend on their proximity to the dome and on the prevailing depositional environment (fig. 3). Potential traps are present wherever sand prevails over mud, and wherever carbonates with enhanced porosity prevail over those with normal porosity.

Pillow State of Growth

This is preserved at the same deep level as turtle-structure anticlines. Pillow growth causes broad crestal uplift so that syndepositionally and post-depositionally thinned strata overlie the pillow crest. If these strata are fluvial, deltaic, or slope-deposited mass-flow deposits, they are sand poor over the pillow crest, but are likely to be flanked by pinch-outs of sandy reservoirs. Sand-rich channels bypass pillow crests and occupy adjacent primary peripheral sinks (fig. 4). Porosity pinch-outs will therefore be found far from the diapir, even near the outer edge of the rim syncline 16 km (10 mi) away. The opposite trend occurs if the pillow grows below shelf siliciclastics or carbonates. Paleotopographic swells over pillow crests are potential reservoirs because they are the preferred sites of reef growth, high-energy grainstone deposition, and sand concentration by winnowing (fig. 5). Here zones of enhanced permeability lie close to the salt stock.

Diapir Stage of Growth

Structural inversion during the onset of diapirism transforms the primary peripheral sink into a turtle-structure anticline cored with sediments coarser than their surroundings. This large, deep trap lies 5 to 20 km (3 to 12 mi) updip of diapirs in the East Texas Basin. During diapirism the marked effects of salt withdrawal beneath the secondary peripheral sink produce stratigraphic traps similar to those in the pillow stage, but are more strongly defined. Comparison with modern depositional environments over active diapirs indicates the same facies relations over the crests of diapirs as over pillows, but over a much smaller area. However, such facies of diapir crests are commonly destroyed by uplift or piercement as the diapir continues to rise. At the diapir stage another important type of trap can form far from the diapir. Between adjacent peripheral sinks are residually high saddles left stranded by the subsidence of salt-withdrawal basins adjacent to them. In a carbonate or shelf environment, both the structure and lithology of these raised areas, which generally overlie much deeper turtle structures, favor the accumulation of hydrocarbons. An example is the giant Fairway field in Lower Cretaceous carbonates in East Texas (fig. 6).

In the Gulf area, diapirs grew during the evolution from an abyssal environment to a shelf environment. The evolution of these diapirs from salt massifs had a major influence on sedimentation patterns. Depotroughs originate in abyssal environments in synclinal depressions between salt massifs that ultimately evolve into clusters of diapirs (figs. 7 and 8). The mass-flow and turbiditic deposits within the depotroughs are most sandy farthest away from the rising salt mounds; the depotroughs lie in channel axes down which the heaviest, coarsest debris moves. Ponds of suitable reservoir sediments can also form immediately updip of the salt massifs (fig. 7). By the time the continental margin progrades to them, the depotroughs have accumulated great thicknesses of sediment because of differential loading of the salt beneath. Some may even have sunk sufficiently deep to rest directly upon the pre-salt substrate, and are therefore stabilized against further local subsidence.

Depopods form on salt massifs when the continental margin has prograded to their area, and therefore tend to be filled with shallow-water deltaic sediments, the load of which causes rapid subsidence and filling, accompanied by vigorous growth of diapirs around them (fig. 8). These reservoirs draped over the crests of rising diapirs characterize embanking offlap units such as the Frio, described later in these notes. These reservoirs make attractive drilling targets and can be recognized bathymetrically by being largely surrounded by diapirs within the boundaries of a salt massif (fig. 7).

Key References: Halbouty (1979), Seni and Jackson (1983a), Woodbury and others (1980)

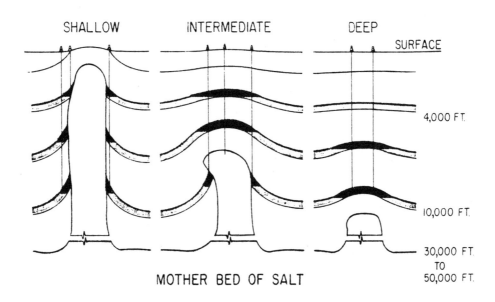

Figure 1. Shallow and intermediate-depth diapirs piercing reservoirs, and a deep, non-piercing diapir. (From Halbouty, 1979.)

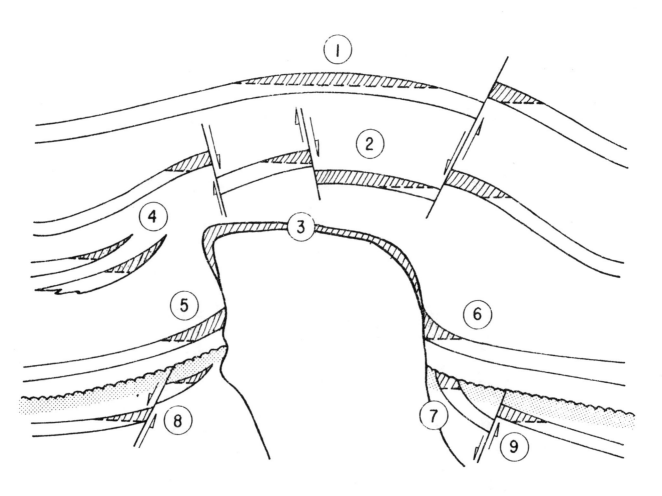

Figure 2. Common types of traps associated with salt domes. 1. anticline; 2. graben caused by extension over salt stock; 3. porous cap rock; 4. flank sand pinch-out; 5. overhang; 6. non-overhanging wall of stock; 7. angular unconformity; 8. normal fault downthrowing reservoir away from dome; 9. normal fault downthrowing reservoir toward dome. (From Halbouty, 1979.)

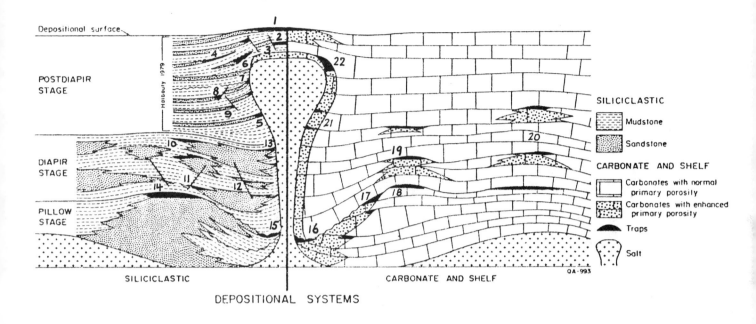

Figure 3. Schematic cross section through a mature diapir showing typical facies variations and potential petroleum traps (numbered) in siliciclastic fluvial, deltaic, and slope depositional systems, and carbonate and siliciclastic shelf depositional systems. Traps 1 through 9 and sand-body geometry in siliciclastic systems deposited during the postdiapir stage are from Halbouty (1979).

1. Combination trap in sand in anticline over crest of dome
2. Graben fault trap over dome
3. Porous cap rock
4. Stratigraphic trap in flank sand pinch-out
5. Structural trap beneath overhang
6. Structural trap uplifted and buttressed against salt stock
7. Unconformity trap
8. Fault trap downthrown away from salt stock
9. Fault trap downthrown toward salt stock
10. Combination trap in sand from updip pinch-out of porous facies in sink
11. Fault trap in sand over turtle structure
12. Fault trap in sand in peripheral sink
13. Stratigraphic trap in sand from domeward pinch-out of porous facies in sink
14. Combination trap in sand at crest of turtle structure
15. Unconformity trap in sand over crest and flanks of precursor pillow
16. Unconformity trap in carbonates from enhanced porosity over crest and flanks of precursor pillow
17. Combination trap in carbonates from pinch-out of enhanced porosity zone on distal flanks of precursor pillow
18. Structural trap in carbonates over crest of turtle structure
19. Combination trap in carbonates from enhanced porosity due to paleotopography over turtle structure
20. Combination trap in carbonates from enhanced porosity due to paleotopography over raised saddle between peripheral sinks
21. Combination trap in carbonates from enhanced porosity near dome due to paleotopography and buttressing against salt stock
22. Combination trap in carbonates over crest of salt stock from enhanced porosity due to paleotopography

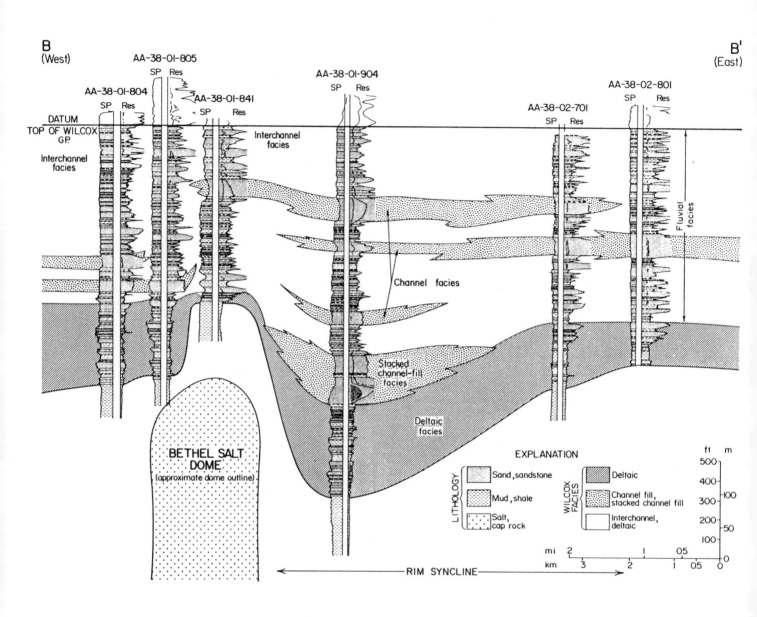

Figure 4. Pinch-out of stacked fluvial channel-fill sands on flanks of Oakwood Dome peripheral sink, East Texas Basin. (From Seni and Jackson, 1983a.)

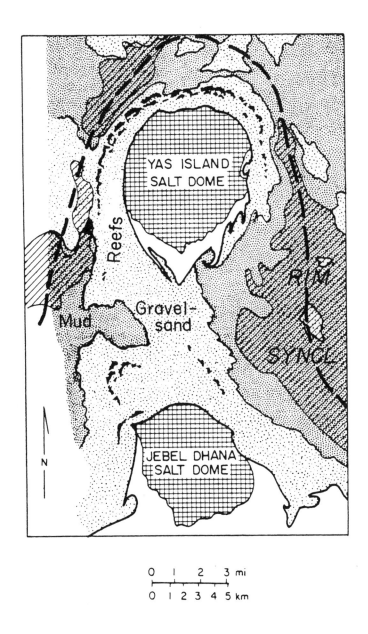

Figure 5. Effect of active diapirism on surrounding facies in a shelf environment of modern carbonate sedimentation in the Persian Gulf, United Arab Emirates. Gravel and sand accumulate in shallow water around the domes by wave winnowing; mud accumulates in the deeper water rim syncline. (Adapted from Purser, 1973.)

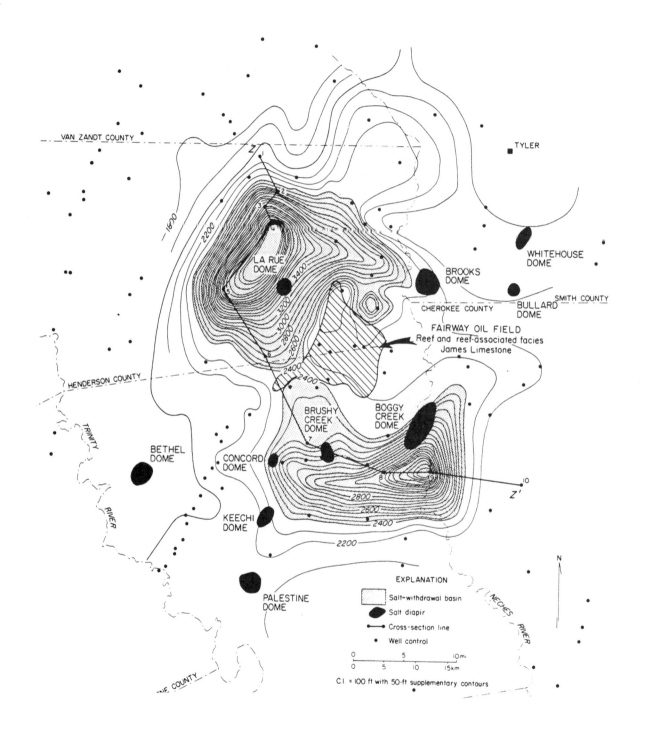

Figure 6. Reef facies, forming preferentially on a residually high saddle between two salt-withdrawal basins, act as a stratigraphic trap for the Fairway oil field. Isopach map, Glen Rose Subgroup, central East Texas Basin. (From Seni and Jackson, 1983a.)

110

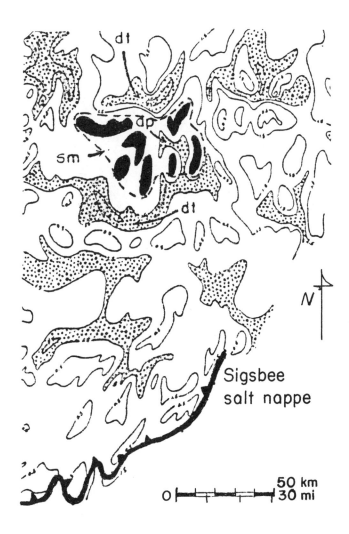

Figure 7. Relations between intraslope basins and salt diapirism on the continental slope, offshore Louisiana. sm, salt massif comprising a cluster of diapirs; dt, depotrough between salt massifs; dp, depopod within salt massif. (Adapted from Spindler, 1977.)

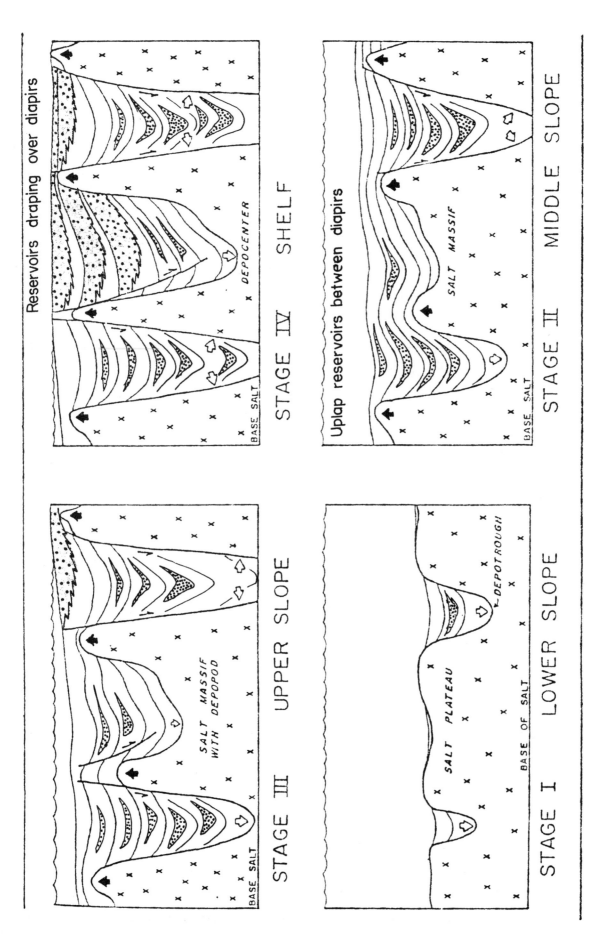

Figure 8. Effects of diapiric evolution on reservoir geometry in a prograding continental margin. Early evolution in depotroughs is characterized by off-structure uplap reservoirs in a mobilizing sedimentary unit. Later evolution is characterized by reservoirs of shallow-water sediments in an embanking unit draped over the structures and concentrated in depopods. (Adapted from Spindler, 1977.)

FAULT TECHNIQUES: ANALYZING NORMAL FAULTS

Where large commercial investments are at stake, it is essential that geologic deductions be as soundly based as possible. For example, many fault traces are drawn on seismic sections solely on the basis of the viewer's perception of the geometry of events that can often be interpreted in more than one way. The geologist should be aware of what fault patterns are realistic, and what are mechanically or geometrically improbable. The following methods and principles assist in:

1. Making geologically realistic interpretations in seismic profiles and cross sections based on well data.
2. Calculating unknown variables such as extension, fault dip, fault throw and shape, and depth to sole out if some of these variables are known.
3. Restoring a cross section to its original (predeformed) state by palinspastic reconstruction.

Two Basic Groups of Extensional Faults

Rotational faults produce extension accompanied by rotation of beds or fault plane. Nonrotational faults produce extension without rotation of beds or fault plane (table 1).

Table 1. Types of normal faults.

Group	Structures rotated	Fault geometry
Non-rotational	Nothing	Planar
Rotational	Beds	Listric
Rotational	Beds and faults	Planar or listric

Nonrotational Extension

In figure 1, extension (*e*) corresponds to heave, and stratigraphic omission (*s*) corresponds to throw if the strata are horizontal. The angle, ϕ, referred to here as the dip of the fault, should be the acute angle between the fault plane and bedding; these two angles are only the same where strata are horizontal. In a series of nonrotational faults, the fault-bed angle, ϕ, is

113

averaged for a whole series. Figure 2 shows how either the stratigraphic omission or the extension can be calculated if the average fault-bed angle is known. For a particular value of s, faults with a low fault-bed angle (particularly less than 5°) have the greatest extension.

Rotational Extension

Planar normal faults: During rotational extension on planar faults (fig. 3) both faults and beds rotate, but the eventual tilt of beds or faults is uniform. Books on a shelf tilt in the same way. The "gaps" (potential voids) are filled by brecciation, cataclasis, or ductile flow. Because beds and faults rotate, calculating extension is different from that in figure 1. Figure 4 can be applied to single faults or to a series of faults with constant bed tilt. Relations between fault dip, bed dip, and extension are shown in figure 5. The most extension is achieved in faults of low dip.

Listric normal faults: Here the curved fault does not rotate during its formation, but the downthrown beds do (fig. 6). The downthrown beds must distort by bedding-plane slip even far from the actual fault plane, as in the right-hand end of the diagram, to maintain contact with the fault plane. Listric faults are always recognizable at surface because of the differential tilt of beds on each side of the fault. If imbricate faults form, they move serially away from the axis of downthrow, except where they are carried by a prograding continental margin. The younger slabs carry and rotate the older ones piggy-back style (fig. 7), and bed tilt increases in the downthrow direction. This rotation can change an antithetic normal fault into a reverse fault (fig. 8). Where a listric master fault bounds a series of synthetic, planar rotational faults (fig. 9), most of the extension takes place by means of the planar faults.

The listric fault model in figure 10 assumes that the fault-bed angle is constant during faulting and that the curved surfaces are arcs of circles. On the basis of figures 4 and 10, we can compare the effectiveness of planar and listric normal faults in extending strata. Figure 11 shows that for a given fault dip and maximum rotation of strata, listric faulting is much less effective than planar faulting.

The potential void can be filled by changing the shape of the downthrown block by minor faulting, as well as by rollover folding. Figure 12 shows the wide range of synthetic and antithetic faults that can form to accommodate the potential void. Two of the reverse faults are primary, and one is a rotated normal fault. It is unreasonable to be biased against interpreting the occasional reverse fault in a zone of overall extension. For instance, in figure 13, the data indicate a reverse fault as strongly as a normal one: both involve repeated sections.

Calculating Depth to Detachment in Listric Faults

During extension by listric normal faulting, the area A in figure 14 is filled by strata displaced from area B, the potential void. Area C has been evacuated by rock to fill the potential void by rollover. The size of area A (which is the same as areas B and C) is a function of h (fault heave) and d (depth to detachment plane or level of sole out). Thus by measuring area C, the depth, d, can be calculated from the heave, h.

Constructing Listric Fault Shape from Rollover Shape

The shape of any part of the rollover reflects the cumulative movement on the listric fault directly below that part (fig. 15). The vector $a'a$ in the fault plane is the result of the throw, t, and the heave, h, for that part of the fault. Similarly the displacement of a' to a on the rollover is also a result of the same throw and heave. Accordingly, if we find the displacement vectors for the rollover, we can project them down to construct the curved fault plane in short segments. Knowing the throw and heave and the rollover shape, construction is as follows:

1. Divide the rollover into blocks with width, h, starting at the fault plane.
2. Draw in the diagonal (plunging _away_ from the fault) to each heave block. This is the throw-heave resultant, or displacement vector.
3. Add these vectors graphically to extend the fault shape downward.

Key References: Gibbs (1983), Wernicke and Burchfiel (1982)

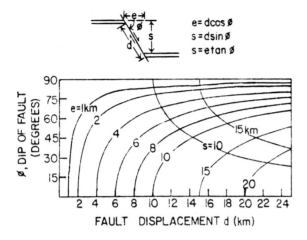

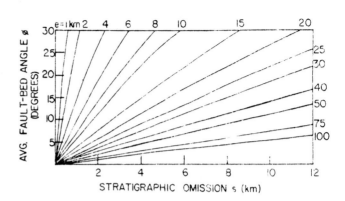

Figure 1. Nonrotational normal faults. Geometry and graphical relations between displacement (d), extension (ε), stratigraphic omission (σ), and fault dip (∅). In horizontal strata, ε is equivalent to heave and σ to throw. (From Wernicke and Burchfiel, 1982.)

Figure 2. Nonrotational normal faults. Plot of stratigraphic omission (σ) versus fault dip or average angle between fault and bedding (∅) for different values of extension (ε). (From Wernicke and Burchfiel, 1982.)

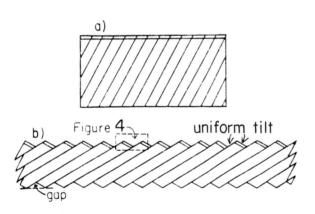

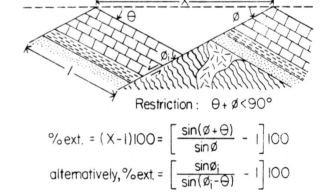

Figure 3. Rotational planar normal faults (bookshelf faulting). (Adapted from Wernicke and Burchfiel, 1982.)

Figure 4. Rotational planar normal faulting. Relation between fault dip (∅) and percentage extension, as proposed by Thompson (1960). (From Wernicke and Burchfiel, 1982.)

116

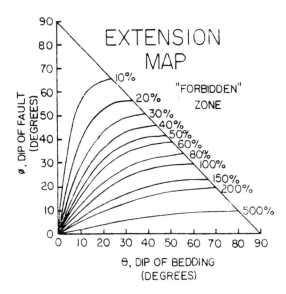

Figure 5. Rotational planar normal faulting. Relations between fault dip (∅), bedding dip (θ), and percentage extension. (From Wernicke and Burchfiel, 1982.)

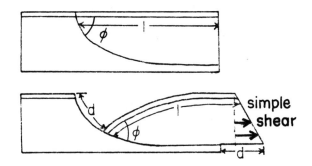

Figure 6. Listric normal fault with reverse drag. Bedding-parallel simple shear (card-deck mechanism) allows the downthrown block to change shape, maintaining contact with the fault plane. (Adapted from Wernicke and Burchfiel, 1982.)

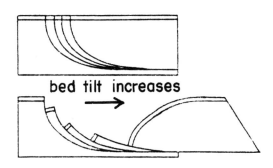

Figure 7. Imbricate listric normal faults. (Adapted from Wernicke and Burchfiel, 1982.)

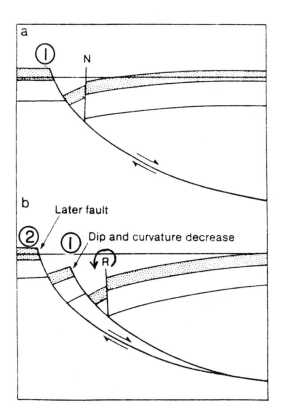

Figure 8. Progressive migration and piggy-back movement during imbricate listric normal faulting. Older fault (1) rotates during later faulting (2); normal antithetic fault (N) is rotated into reverse fault (R) geometry. (From Gibbs, 1983.)

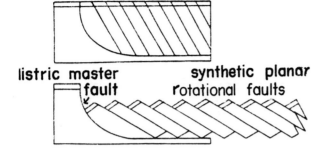

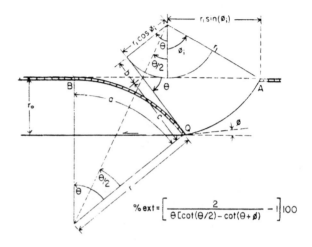

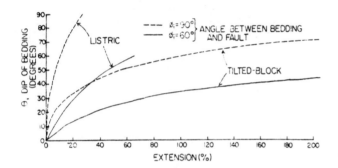

Figure 9. Listric normal master fault bounding a family of synthetic planar rotational faults. (Adapted from Wernicke and Burchfiel, 1982.)

$$\% \, ext = \left[\frac{2}{\theta \left[\cot(\theta/2) - \cot(\theta + \phi) \right]} - 1 \right] 100$$

Figure 10. Listric fault model showing relations between angle between bedding and fault (ϕ), maximum dip of rollover bedding with respect to regional bedding (θ), and extension (AB - a). Model assumes that ϕ remains constant during faulting and that curvilinear segments are arcs of circles. (From Wernicke and Burchfiel, 1982.)

Figure 11. Planar rotational normal faults (tilted block or bookshelf mechanism) yield greater extension than do listric normal faults for a given maximum stratal rotation (θ) and fault-bedding angle (ϕ). (From Wernicke and Burchfiel, 1982.)

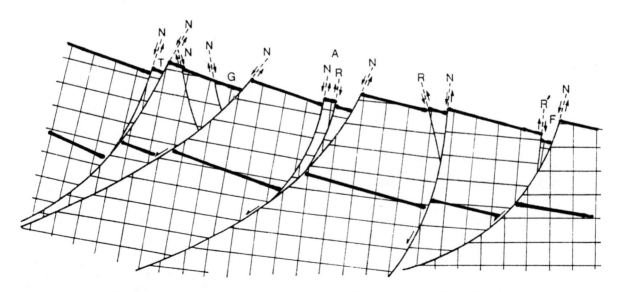

Figure 12. Fault structures forming during extension without rollover folding. N, normal fault; R, reverse fault; R', rotated normal fault; A, horst; G, graben; F, horst-foot graben; T, crestal terrace. (Adapted from Gibbs, 1983.)

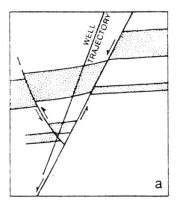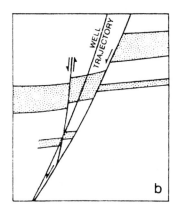

Figure 13. Problems interpreting synthetic faults. Both a reverse fault (a) and a normal fault (b) yield repeated section and cannot be differentiated in a single well. However, the latter only arises where boreholes are nonvertical. (From Gibbs, 1983.)

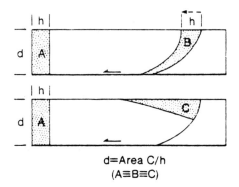

$$d = \text{Area } C/h$$
$$(A \equiv B \equiv C)$$

Figure 14. Area B, the potential void, is closed by accommodation structures during faulting. The equivalent area A is a product of d, the depth to detachment where the master fault soles out, and h, the fault heave. Thus, d can be calculated by measuring h and area c, assuming cross-sectional area is constant during faulting. (From Gibbs, 1983.)

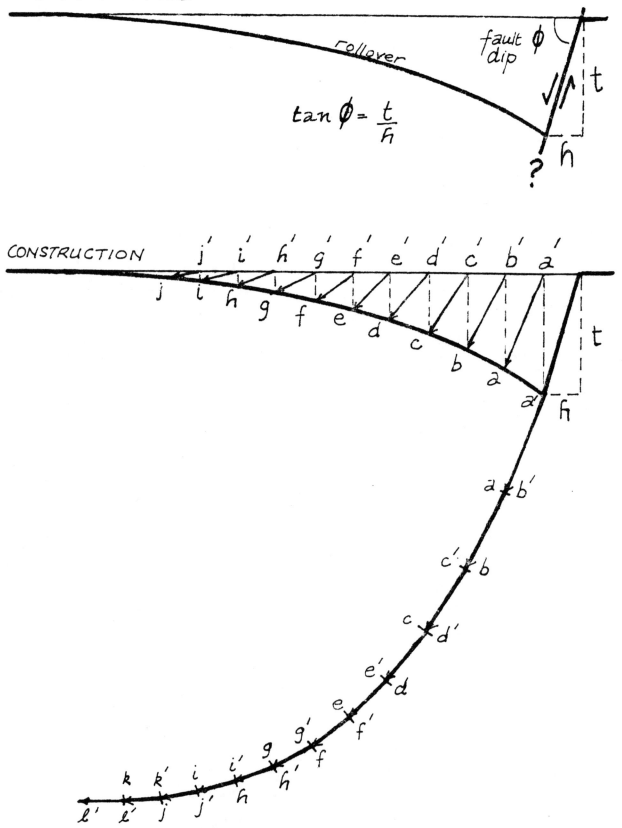

DATA SUPPLIED

rollover

fault dip ϕ

$\tan \phi = \dfrac{t}{h}$

CONSTRUCTION

Figure 15. Constructing listric fault shape at depth from the shape of the overlying rollover fold.

FAULT TECHNIQUES: BALANCING CROSS SECTIONS WITH EXTENDED STRATA

A geologic cross section should integrate all known borehole, geophysical, and surface data; information off the line of section should be projected along strike (along plunge in folded strata) into the section. A cross section that is balanced is not necessarily correct, but one that is not balanced is invariably wrong unless a valid explanation is provided in the cross section. Structural balancing not only produces a more accurate cross section, but it also tests the ideas that we build into the section.

To be balanced, a cross section should be drawn (1) without vertical exaggeration, and (2) parallel to the direction of maximum regional extension (perpendicular to the strike of normal faults). Constant cross-sectional area before and after deformation must be assumed unless strain data are obtainable from the actual rocks.

Basic Principles

Figure 1A shows the simple principle behind balancing extended strata, where the original length is l_0 and the extended length is l_1. Because cross-sectional area is assumed to be the same before and after strain, area A (which is lost by thinning) must equal area B (which is gained by extension). This is the same principle that allows us to calculate depth to detachment in the previous section on listric faults, but it can be applied to any type of deformation in which constant cross-sectional area is preserved. By measuring area A and calculating the extension (e), we can calculate the average depth to detachment of an irregularly extended terrane. Area A should include only fault-induced subsidence, not flexural subsidence (fig. 1B). The original elevation of the marked horizon before faulting can be estimated by extrapolating the unfaulted horizon across the faulted area, keeping it parallel to the bedding (fig. 1B). This cannot be done where listric normal faults have rotated the bedding (fig. 1C); here palinspastic reconstruction can be attempted by rotating the fault blocks back to their unfaulted positions and orientations so that the original elevation of the marker horizon can be estimated.

Displacements on individual faults must also be balanced. If separate boreholes indicate differential displacement at different places on the same fault, the section must account for this. Differential displacement can be compensated for by horsetailing of the fault.

Oblique Cross Sections

To convert data where the section is α degrees from the maximum extension direction:

$$l_\lambda = (l_c{}^2 - (S . \sin\alpha)^2)^{1/2}$$

$$\theta_\lambda = \sin^{-1}(S . \sin\alpha / l_\lambda)$$

Where:

l_λ = length of true section parallel to extension direction

l_c = length of observed segment over horizontal distance, S

θ_λ = dip of l_λ

Effects of Compaction

Apart from actual faulting, compaction is the most important element of strain in a diverging continental margin.

Estimate compaction by

1. Direct measurement of deformed objects (e.g., reduction spots, burrows) in core. These estimates only apply to discrete intervals containing the objects.
2. Use general compaction curves for depth of burial. As this depends on burial rate, general curves may not be representative of the area under study.
3. Estimate from sonic logs.

Remove effects of compaction:

During palinspastic reconstruction, as each overlying layer is "backstripped" (removed), underlying units can be decompacted to their former states when they lay at shallower depths. Decompaction not only makes sedimentary units thicker, but it also makes faults steeper, changing their throw but not their heave. Diapiric effects should also be removed.

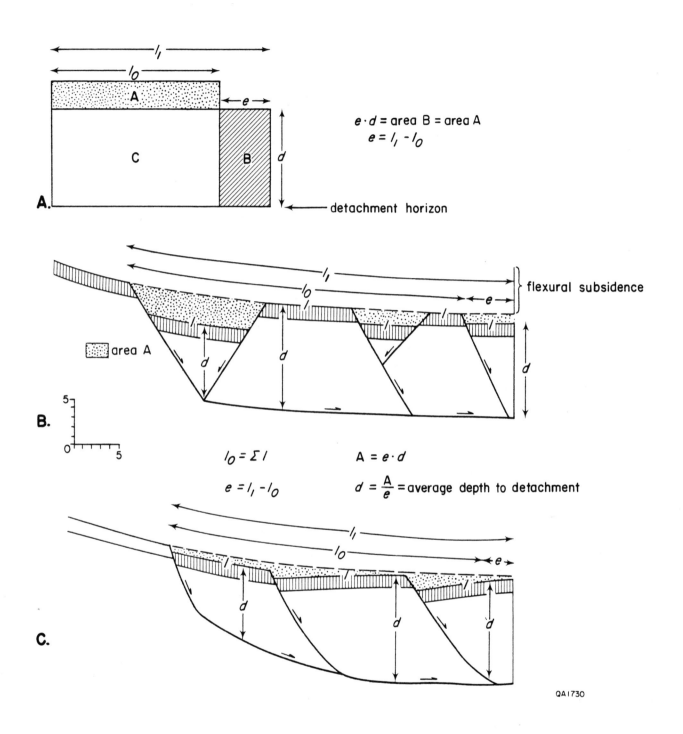

Figure 1. Area balance and calculation of depth to detachment for extended terranes, assuming constant cross-sectional area during extension. A. Basic principles. B. Simplified extension by nonrotational planar faulting. C. More realistic extension by listric normal faulting. Area A subsidence is entirely caused by faulting; flexural subsidence must not be included in this area. (Adapted from Gibbs, 1983.)

SALT-DOME TECHNIQUES: ISOMETRIC CONSTRUCTION OF THREE-DIMENSIONAL SHAPES

<u>Requirements:</u> structure-contour map

<u>Key Reference:</u> Lobeck (1924)

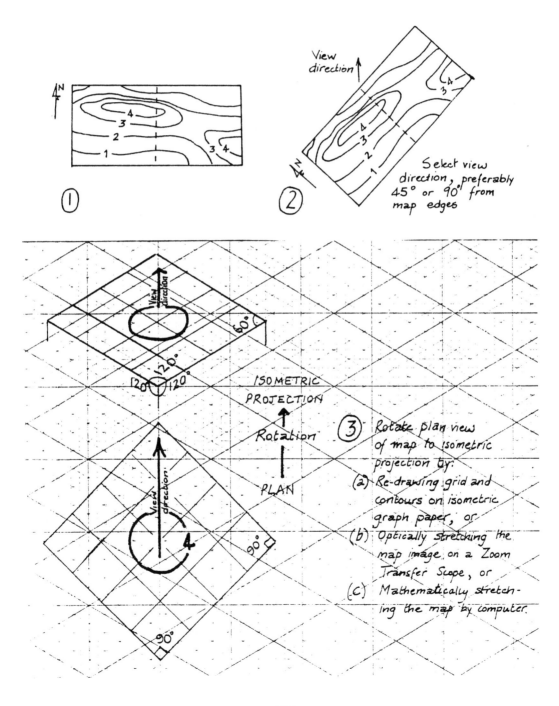

①

② Select view direction, preferably 45° or 90° from map edges

ISOMETRIC PROJECTION

Rotation

PLAN

③ Rotate plan view of map to isometric projection by:

(a) Re-drawing grid and contours on isometric graph paper, or

(b) Optically stretching the map image on a Zoom Transfer Scope, or

(c) Mathematically stretching the map by computer.

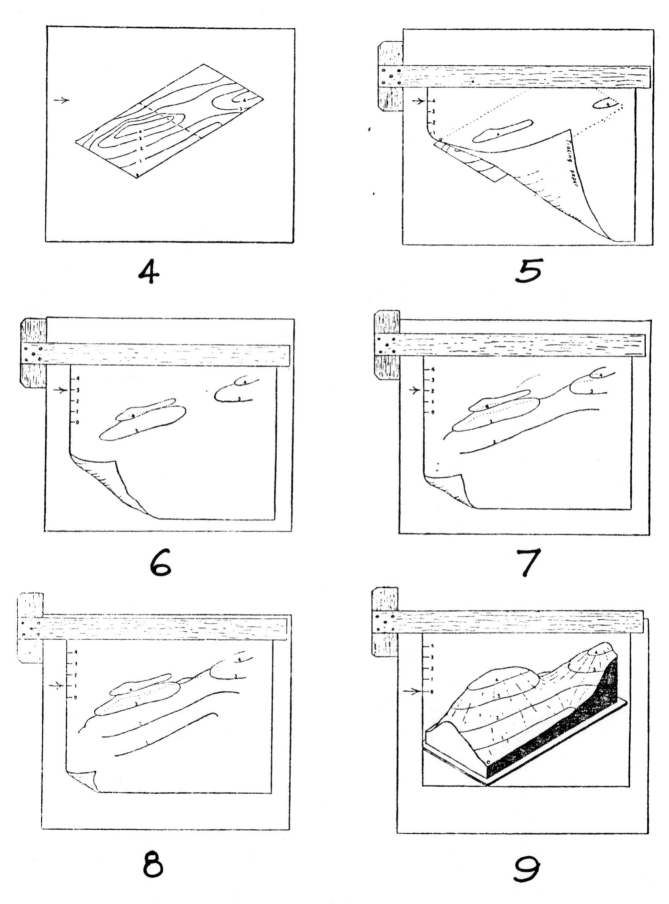

4

5

6

7

8

9

126

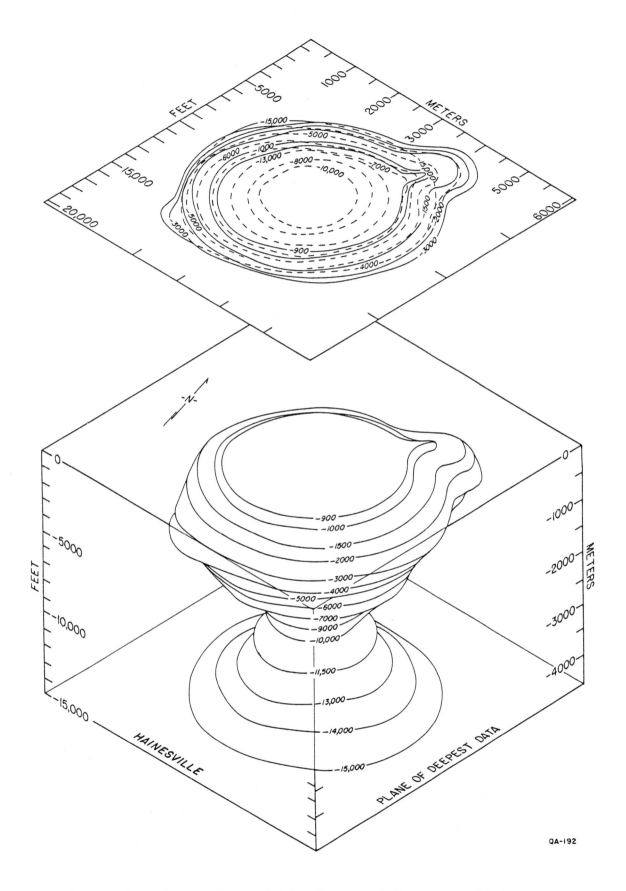

Figure 1. Isometric block diagram of Hainesville salt stock (East Texas Basin), drawn by Lobeck's (1924) technique with no vertical exaggeration. (From Jackson and Seni, 1984a.)

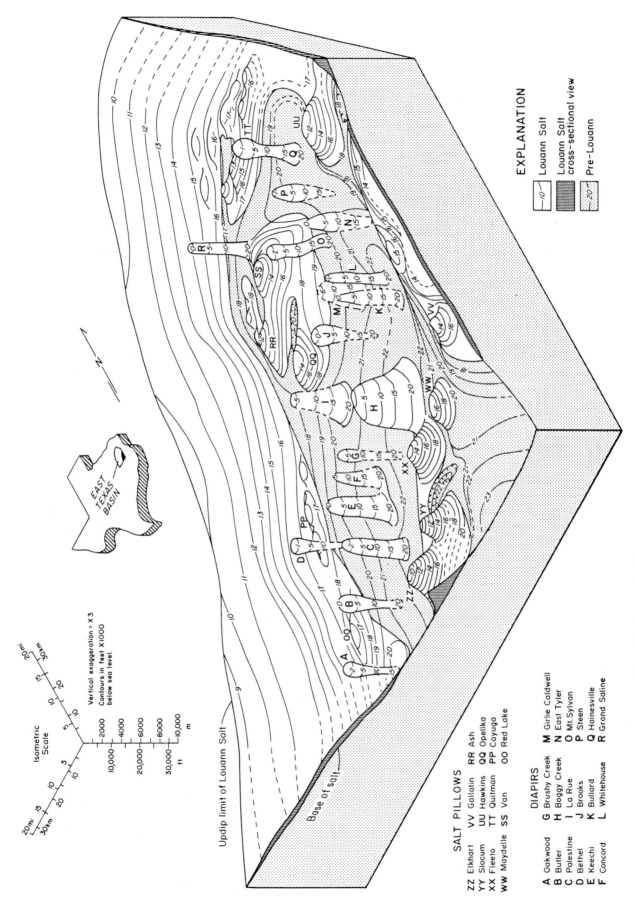

SALT PILLOWS

ZZ Elkhart   VV Gallatin   RR Ash
YY Slocum   UU Hawkins   QQ Opelika
XX Fleeto   TT Quitman   PP Cayuga
WW Maydelle   SS Van   OO Red Lake

DIAPIRS

A Oakwood   G Brushy Creek   M Girlie Caldwell
B Butler   H Boggy Creek   N East Tyler
C Palestine   I La Rue   O Mt. Sylvan
D Bethel   J Brooks   P Steen
E Keechi   K Bullard   Q Hainesville
F Concord   L Whitehouse   R Grand Saline

EXPLANATION

Louann Salt

Louann Salt
cross-sectional view

Pre-Louann

Figure 2. Isometric block diagram of the East Texas Basin salt structures, drawn by Lobeck's (1924) technique with vertical exaggeration. (From Jackson and Seni, 1983.)

128

SALT-DOME TECHNIQUES: RECONSTRUCTING SALT FLOW

Reconstructing Dome Growth in Cross Section

Requirements: (1) Migrated seismic profile, converted to depth, true scale (fig. 1)

(2) Compaction estimates for all intervals, based on density measurements in core, sonic logs, velocity surveys, or compaction-depth curves

Procedure (fig. 1):

Add each unit bounded by seismic reflections on top of the previous one, keeping the upper reflector horizontal at the surface. Decompact each unit to its original thickness before placing on the previous unit (this has not been done in fig. 1). As each unit is progressively buried during each increment of reconstruction, compact it proportionally to its depth at that stage.

Reconstructing Dome Growth on Maps

Requirements: (1) Known distribution of salt pillows and diapirs

(2) Structure-contour map of a post-salt horizon at moderately deep level, commensurate with good subsurface control (fig. 2)

Procedure:

1.   Delineate all axial traces on the structure-contour map. Everywhere an axial trace of a fold changes its direction of plunge, that fold is crossfolded by another; the axial traces of both folds should be marked on the map (fig. 3). During the procedure, continually rearrange the map orientation so that strata dip away from you. Folds will then plunge away from you and anticlines will be arch-shaped and synclines will be trough-shaped for easy recognition.

2.   Interpret the map to locate the positions of original, prediapir, salt paleoridges, based on the following two principles (fig. 4). (a) Pillows lie on regional anticlines, but if the pillows evolve to diapirs, their deflation causes structural inversion and the diapirs are overlain by regional synclines (fig. 5). Thus the regional synclines over diapirs mark the

traces of the original salt anticlines. So join up regional anticlines over pillows with regional synclines above diapirs. (b) Upward growth of a diapir forms a local dome that distorts the regional syncline overlying the diapir. The local dome either deflects the syncline to one side (e.g., East Tyler and Boggy Creek Domes in fig. 3), or it splits the regional syncline into a rim syncline around the dome (e.g., Hainesville and Mount Sylvan Domes in fig. 3). To remove this dome-induced distortion, draw the reconstructed salt paleoridges through the center of each diapir (and pillow). Each diapir will have grown from the crest of a salt paleoridge or pillow and will therefore mark the position of that crest.

Key Reference: Loocke (1978)

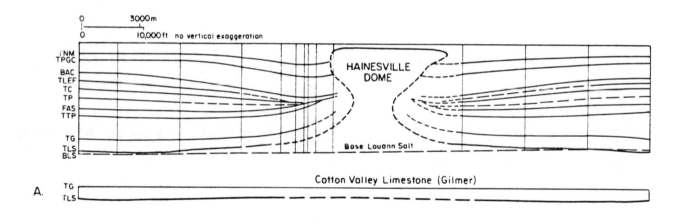

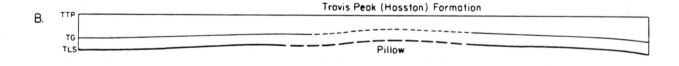

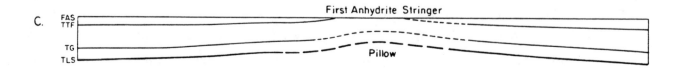

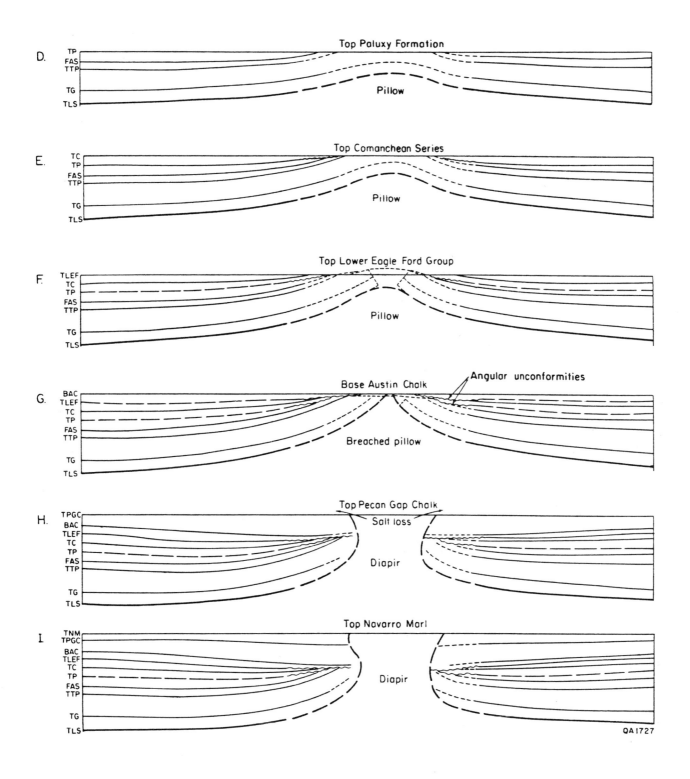

Figure 1. Palinspastic growth reconstruction of Hainesville Dome (East Texas Basin) in cross section, based on the seismic reflectors shown in the top diagram; strata have not been decompacted to their original thicknesses, which is required for quantitative reconstruction. (Adapted from Loocke, 1978.)

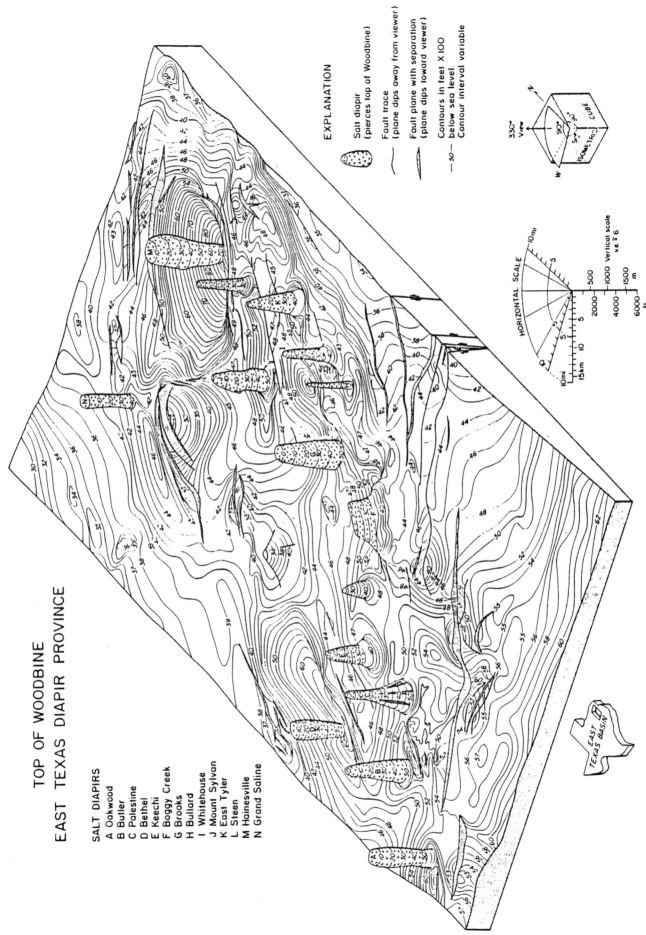

TOP OF WOODBINE
EAST TEXAS DIAPIR PROVINCE

SALT DIAPIRS

A Oakwood
B Butler
C Palestine
D Bethel
E Keechi
F Boggy Creek
G Brooks
H Bullard
I Whitehouse
J Mount Sylvan
K East Tyler
L Steen
M Hainesville
N Grand Saline

EXPLANATION

Salt diapir
(pierces top of Woodbine)

Fault trace
(plane dips away from viewer)

Fault plane with separation
(plane dips toward viewer)

—50— Contours in feet X 100
below sea level.
Contour interval variable

Figure 2. Isometric block diagram of salt diapirs and structure contours on top of the Woodbine Group, East Texas Basin. (From Jackson and Seni, 1984b).

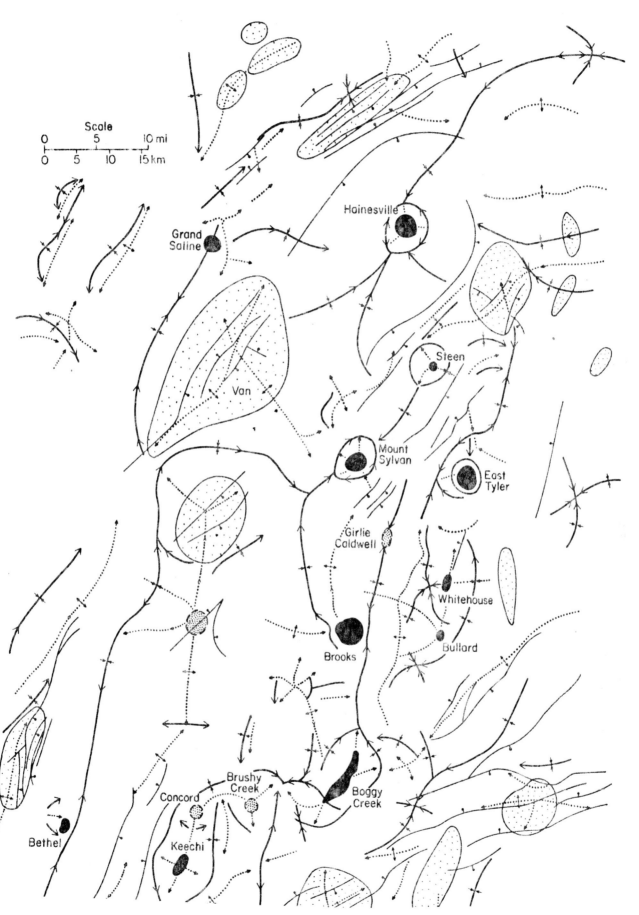

Figure 3. Fold axial traces on the top of the Woodbine Group, interpreted from the structure contours shown in figure 2. Black, piercing diapirs; heavy stipple, nonpiercing diapirs; light stipple, salt pillows.

133

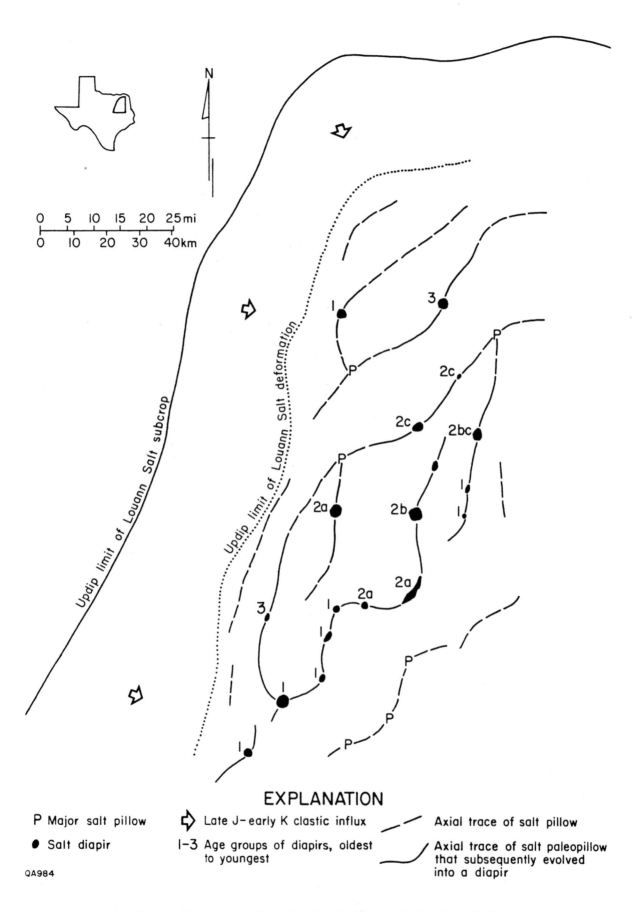

QA984

EXPLANATION

P Major salt pillow   ⇨ Late J−early K clastic influx   – – Axial trace of salt pillow

● Salt diapir   1−3 Age groups of diapirs, oldest to youngest   ⌢ Axial trace of salt paleopillow that subsequently evolved into a diapir

Figure 4. Reconstructed Jurassic salt paleoridges in the East Texas Basin.

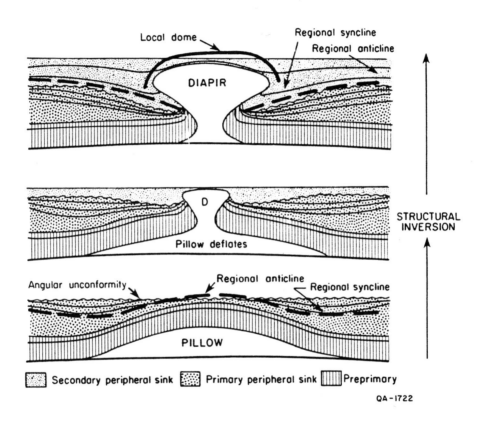

Figure 5. Structural inversion of regional folds during evolution of a salt pillow to a salt diapir. (Adapted from Trusheim, 1960.)

SALT-DOME TECHNIQUES: ESTIMATING DOME GROWTH RATES

To understand the geologic history of strata around a salt dome, we need to consider the rate of rise of the salt relative to the surrounding strata--the dome growth rate--during all three stages of dome evolution (pillow to diapir to postdiapir stages).

Types of Growth Rate

Gross rates are a function of the volume of salt evacuated from the withdrawal basin and mobilized up the diapir. Net rates are a function not only of this process, but of all other processes that affect diapir height and growth rate, such as salt dissolution, extrusion, and lateral intrusion. Thus gross rates of growth approximate the true rate of salt flow within the diapir, regardless of the independent motion of the actual diapir crest. On the other hand, net rates of growth approximate the actual movement of the diapir crest, which is of greater concern to petroleum geologists.

Calculating Growth Rates

The stratigraphic section is divided into broad units (preferably easily picked in well logs and seismic sections, and of known duration). Rates are calculated separately for each unit. These yield a composite growth curve for each dome when combined.

1. The volume of sedimentary fill in a salt-withdrawal basin equals the volume of salt that migrated during filling of that basin (fig. 1). This volume divided by the duration of withdrawal quantifies the gross rate of salt movement into a diapir in terms of volume per time.

2. Net rates of salt-pillow uplift are calculated by the rate of stratigraphic thinning per time over the crest of salt pillows, which are nonpiercing (fig. 2A). Net rate is obtained by dividing growth by duration.

3. Net rates of diapir crest uplift are equated with the maximum rate of sediment accumulation in the associated withdrawal basin (fig. 2B). This technique assumes that the salt remained near surface throughout most of the dome growth history. Upward movement of salt was balanced by basin subsidence, ground-water dissolution, and extrusion. Again, rate is obtained by dividing growth by duration.

4.   Gross rates of diapir elongation are calculated by dividing the volume of salt moved (estimated by method 1) by the maximum horizontal cross-sectional area of the diapir (fig. 2C). This technique assumes that all salt migrated from below the withdrawal basin into the diapir and rose through a constriction represented by the cross-sectional area of the salt stock, thereby lengthening the stock. Rate is obtained by dividing growth by duration.

Key References: Seni and Jackson (1983a,b)

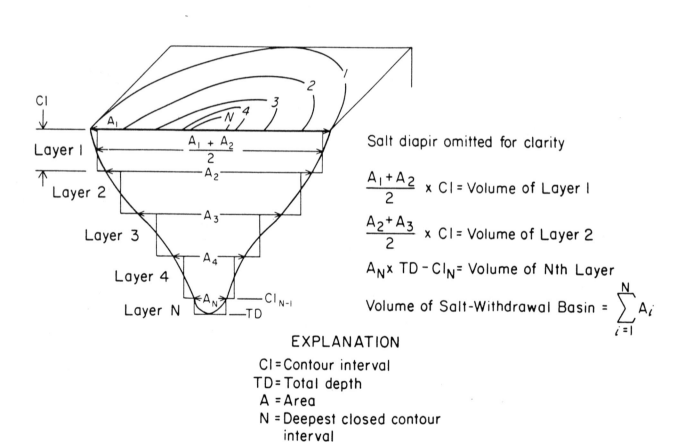

Salt diapir omitted for clarity

$$\frac{A_1 + A_2}{2} \times CI = \text{Volume of Layer I}$$

$$\frac{A_2 + A_3}{2} \times CI = \text{Volume of Layer 2}$$

$$A_N \times TD - CI_N = \text{Volume of Nth Layer}$$

$$\text{Volume of Salt-Withdrawal Basin} = \sum_{i=1}^{N} A_i$$

EXPLANATION

CI = Contour interval
TD = Total depth
A = Area
N = Deepest closed contour
       interval

Figure 1. Technique for calculating volume of an individual salt-withdrawal basin. (From Seni and Jackson, 1983b.)

Method

Application, Assumptions, Restriction, Advantages

A. Net Growth by Sediment Thinning

Not to scale

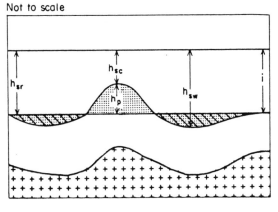

$$h_p = h_{sr} - h_{sc}$$
$$G_n = h_p$$
$$\dot{G}_n = \frac{h_p}{t_i}$$

h_p = Height of pillow

h_{sr} = Regional mean sediment thickness

h_{sc} = Minimum sediment thickness over crest of structure

h_{sw} = Maximum sediment thickness in salt-withdrawal basin

G_n = Net growth of pillow

$\dot{G}_n$ = Net rate of growth of pillow

t_i = Duration of stratigraphic interval (i)

Net growth calculated by sediment thinning will equal net growth calculated by sediment thickening only when

$$h_{sc} = 0$$
$$\text{and } h_{sw} = h_{sr}$$
$$\therefore G_n = h_{sr} - h_{sc} = h_{sw}$$

Application: Pillow stage, postdiapir stage (only for non-pierced strata)

Assumptions:

(1) Sediment thinning is syndepositional

(2) Sediment thinning is due to uplift of crest of structure

Restriction: Only records extension greater than shortening caused by extrusion or dissolution

Advantages:

(1) Simple quantitative methodology

(2) Can be measured from single cross section

(3) Applicable to youngest strata not pierced by diapir, thus provides rates of most recent growth

Figure 2. Methods for calculating net rate of pillow growth (A), and net (B) and gross (C) rates of diapir growth, and the applications, assumptions, restrictions, and advantages of each method. (From Seni and Jackson, 1983b.)

Method	Application, Assumptions, Restriction, Advantages

B. Net Growth by Sediment Thickening

Not to scale

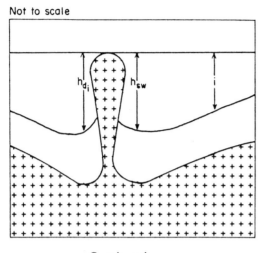

$$G_n = h_{d_i} = h_{sw}$$

$$\dot{G}_n = \frac{h_{sw}}{t_i}$$

h_{d_i} = Height of diapir

h_{sw} = Maximum thickness of stratigraphic interval measured in salt-withdrawal basin

G_n = Net growth of diapir

$\dot{G}_n$ = Net rate of growth of diapir

t_i = Duration of stratigraphic interval (i)

Application: Pillow stage, diapir stage, postdiapir stage

Assumptions:

(1) Diapir remains near sediment surface during deposition

(2) Rate of deposition controls or is controlled by diapir growth

Restriction: Only records net extension greater than shortening caused by diapir extrusion or dissolution

Advantages:

(1) Simple quantitative methodology

(2) Can be measured from single cross section

Method	Application, Assumptions, Restrictions, Advantage

C. Gross Diapir Growth

Not to scale

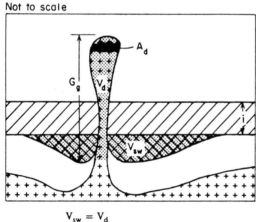

$$V_{sw} = V_d$$

$$G_g = \frac{V_{sw}}{A_d}$$

$$\dot{G}_g = \frac{V_{sw}/A_c}{t_i}$$

V_{sw} = Volume of sediments in salt-withdrawal basin

V_d = Volume of diapir

A_d = Maximum cross-sectional area of diapir

G_g = Gross growth of diapir

$\dot{G}_g$ = Gross rate of growth of diapir

t_i = Duration of stratigraphic interval (i)

Application: Diapir stage only

Assumptions:

(1) Present cross-sectional area of diapir equals cross-sectional area of diapir during filling of withdrawal basin

(2) Volume of withdrawal basin equals volume of salt mobilized during filling of withdrawal basin

(3) All salt mobilized during filling of withdrawal basin migrated into diapir

Restrictions:

(1) Requires measuring volume of withdrawal basin and area of diapir, which requires close well spacing for map construction

(2) Growth by tear-drop detachment of diapir base is not measurable

Advantage: Records total extension independent of possible dissolution or extrusion

TERTIARY GULF COAST REGIONAL SUBSIDENCE

Types of Regional Subsidence

Tectonic development and gross structure are controlled by enormous subsidence of the edge of a diverging continental margin like the Gulf Coast. This subsidence is primarily a response to extending lithosphere.

Rift subsidence operates before continental breakup and possibly persists. It is generally followed by flexural subsidence, or downwarping of the margin of the basin without obvious fault control (fig. 1). In dealing with the Tertiary Gulf Coast we are not concerned with the rift stage but with the flexural phase of regional subsidence.

The mechanics of passive-margin subsidence continues to be intensively studied and we can only scratch the surface of current hypotheses. A major unknown is whether the lithosphere deforms as an elastic substance like rubber or as a viscoelastic substance like putty. The three principal mechanisms driving regional flexural subsidence are loading, cooling, and thinning (fig. 2).

Loading Model

Loading can create a basin if the load is a volcanic pile, ice sheet, or thrust sheet. Sediment loading alone cannot create a basin, although it can greatly modify existing basins.

Consider the thickness, h, of sediment that can accumulate by classic (Airy) isostasy at a continental margin, with an initial water depth, d

$$h = d \, (\rho_m - \rho_w)/(\rho_m - \rho_s)$$

where ρ_m = mantle density = 3.30 g/cm^3
ρ_w = seawater density = 1.03 g/cm^3

With average sediment densities (ρ_s) of 2.15 and 2.55 g/cm^3, subsidence is approximately $2d$ and $3d$, respectively. Thus in shelf conditions, where d is less than 200 m, loading alone can only produce subsidence of up to 600 m. However, if sediment loads the continental slope, originally 2,000 m deep, this load can cause many kilometers of subsidence beneath a depocenter (fig. 3b). More realistically, assuming that the crust flexes elastically under load, the subsidence zone is extended landward and seaward by more than 100 km (fig. 3c).

The importance of sediment loading is that it accentuates the regional subsidence caused by cooling or thinning of the crust. Load-induced subsidence has allowed the formation of fat depocenters at the margin, where shallow-water sediments overlie thick deep-water prodelta muds.

Thermal Model

This hypothesis assumes that the continental lithosphere near a splitting margin is heated. Heating is a natural result of stretching (fig. 4A) because isotherms are drawn upward into the thinning zone, steepening the geothermal gradient. Heating expands the crust, uplifting and exposing it to erosional thinning. Two km of uplift and erosion will be followed by about 2 km of isostatic subsidence of the continental margin. This subsidence can be augmented by subsidence due to density increases of the crust caused by metamorphism or injection of mafic dikes; this yields about 3 or 4 km more subsidence.

As the thinned lithosphere subsequently cools, the underlying asthenosphere freezes, thickening the lithosphere and increasing its strength to greater than its original state (fig. 5).

Stretching Model

Stretching alone can cause subsidence of continental crust. If the crust/lithosphere ratio is more than 0.1375 (corresponding to a crust thickness of more than 17 km if the lithosphere is 125 km thick; average continental crust isostatically stable at sea level is 31.5 km), stretching causes subsidence, S_i. The thermal effects described above also come into play, causing a combined subsidence, S_t. Total subsidence, S, is subsidence due to stretching, cooling, and sediment loading.

Figure 6 shows the combined effects of initial stretching and later cooling of the thinned crust and underlying lithosphere mantle; also shown (right-hand ordinate) are the effects of superimposed sediment loading if the basin is completely filled by sediments throughout subsidence. β is the ratio between the original length (say 1 unit) and the stretched length (β) of the crust. The crust is thinned by a factor of $1/\beta$. If thinning proceeds to the point of continental splitting and ocean-floor formation ($\beta = \infty$), the maximum subsidence for crust and lithosphere of thickness 31.2 and 125 km is theoretically 3.4 km. But we observe that the present mean depth of oceanic ridges (this depth being equivalent to S_i for $\beta = \infty$) is 2.6 km. Thus the continental Pangean crust that originally split to form the ridges could not have been more than 28.2 km thick.

The combination of stretching, cooling, and loading has allowed most of the Gulf of Mexico to subside to about 11 km below sea level. Oceanic crust in the center of the Gulf indicates that here $\beta = \infty$. Below the Cenozoic depocenter above thinned continental crust, β is likely to be between 1.5 and 4.0. A value of $\beta = 2.7$ is compatible with the total subsidence of the Gulf since its formation. Using this value, rates of total subsidence in the Early Tertiary and today can be estimated at 25 m/Ma and 20 m/Ma, respectively (fig. 6). These are considerably slower rates than the average Cenozoic rate of sediment accumulation, which is about 120 m/Ma. Again, using the same value of β, total subsidence throughout the 65-Ma Cenozoic can be estimated at 1.2 km (fig. 6). In order for 6.5 km of observed Cenozoic sediment to accumulate, water depth at the start of the Tertiary must have been at least 5.3 km, which is deeper than the present abyssal plain in the center of the Gulf. Clearly, the Gulf margins must have been deep and sediment starved at the end of the Mesozoic in order to trap the great sediment thickness (fig. 7); the Gulf center is still starved.

Summary

The enormous subsidence and thick sediment package at the edge of the Gulf continental margin is the result of the following.

1. Crustal thinning and cooling created a starved basin subsiding faster than it could be filled in the Mesozoic. Thus the rate of progradation of the continental margin was slow relative to the large volumes of detritus received. Facies were stacked vertically, magnifying the effects of sediment loading.

2. Sediment loading of the crust allowed isostatic subsidence to depress the crust deeply beneath the prograding depocenter. Loading was especially effective because the basin was starved and the foundation of the prograding margin was already very deep.

Key Reference: Bott (1980)

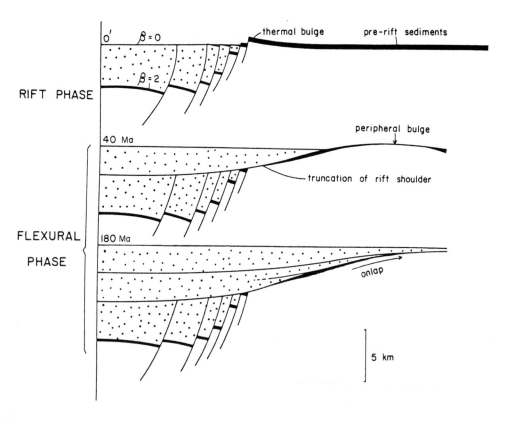

Figure 1. Evolution of a sedimentary basin during rift and flexural phases assuming the lithosphere stretching model of McKenzie (1978). (From Dewey, 1982.)

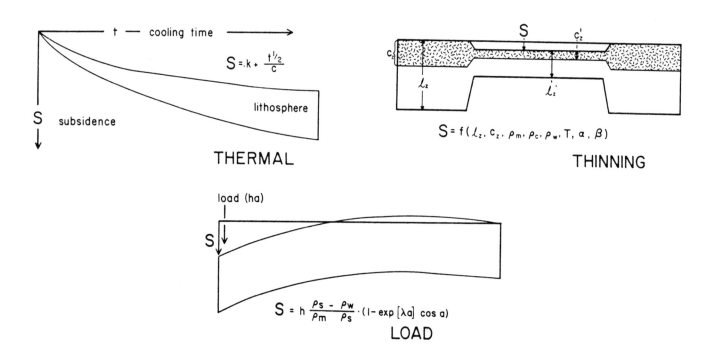

Figure 2. Three principal mechanisms of basin subsidence (S): thermal due to cooling and thickening of the lithosphere; thinning due to stretching of the crust (thickness c_z) and lithosphere (thickness l_z); and loading due to sedimentary fill. (From Dewey, 1982.)

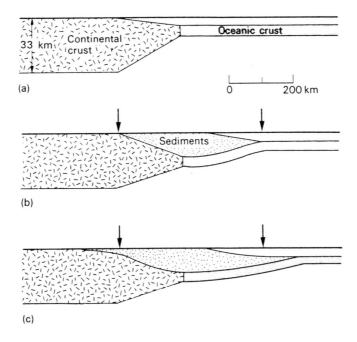

Figure 3. Sedimentary loading. A. Before loading, a 200-km-wide transition zone separates continental crust from oceanic crust. B. Simple Airy loading by sediment fill, assuming density of sediments is 2.45 g/cm^3 and that of upper mantle is 3.30 g/cm^3. C. Equivalent loading, but assuming flexure of an elastic lithosphere with flexural rigidity of 2 x 10^{22} N.m. (From Bott, 1979.)

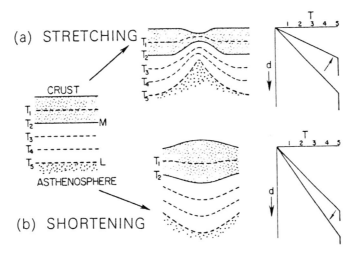

Figure 4. Behavior of compositional boundaries and isotherms (T_1 through T_5) during stretching and shortening of the lithosphere (crust ornamented). M, Mohorovicic discontinuity; L, base of lithosphere; d, depth. (From Oxburgh, 1982.)

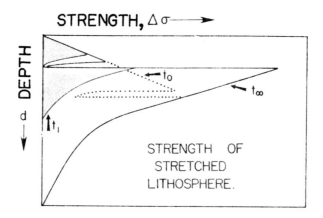

Figure 5. Strength of continental lithosphere; prestretching (t_0 dotted); immediately after stretching (t_1 toned profile); after cooling to equilibrium (t_∞ solid line). In all profiles the strength minimum corresponds to the base of the crust. (Adapted from Oxburgh, 1982.)

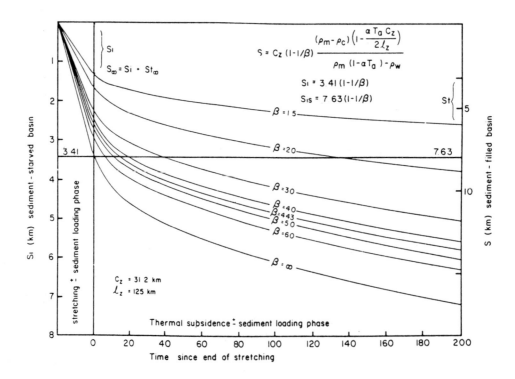

Figure 6. Subsidence during uniform lithospheric stretching (S_i) and thermal recovery (S_t) for initial crustal and lithosphere thicknesses of 31.2 and 125 km, respectively, for β values of 1.5 to ∞. Left ordinate shows only subsidence due to stretching and cooling; right ordinate shows total subsidence (S), including effects of sediment loading assuming basin remains full. (From Dewey, 1982.)

146

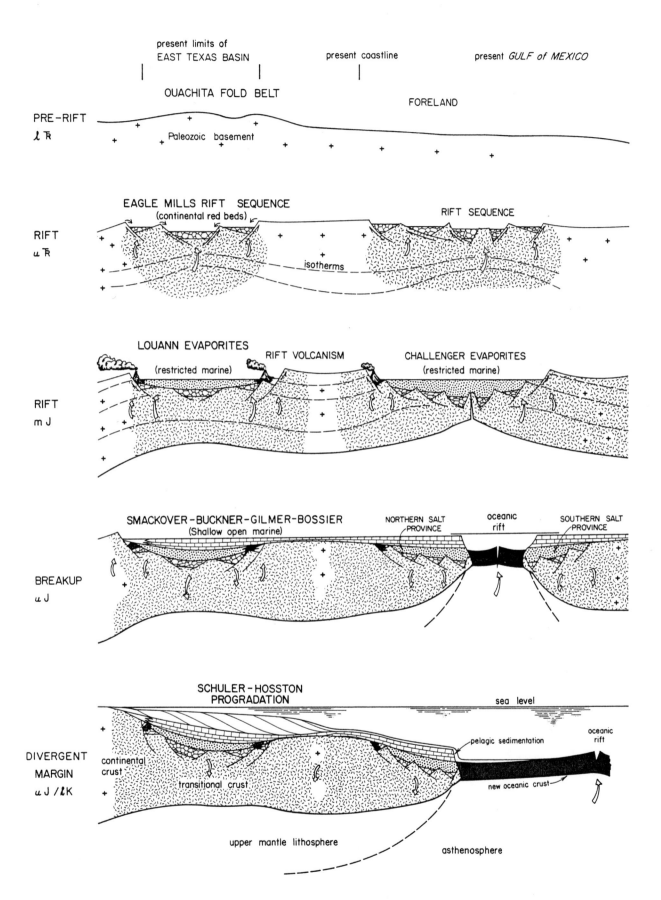

Figure 7. Schematic northwest-southeast cross sections showing evolutionary stages in formation of the northern Gulf of Mexico and East Texas Basin. Arrows indicate thermally induced isostatic movement of crust. (From Jackson and Seni, 1983.)

GULF COAST TERTIARY STRAIN DOMAINS

Enhanced Local Subsidence

Progradational clastic continental margins like the northern Gulf of Mexico can be divided into stable (e.g., Woodbine and Midway margins) and unstable (e.g., Tuscaloosa and Wilcox margins) types. In the unstable type, large-scale normal faulting at the margin modifies the original depositional geometry and allows local subsidence rates much greater than the regional subsidence (fig. 1). Sediments accumulating here can thus form a substantial part of the total basin fill (fig. 2). The growth faults individually propagate upward and their collective activity shifts basinward with the prograding continental margin.

Strain Domains

The local subsidence at the continental margin allowed by normal faults has a corresponding extensional component. Sediments slide downdip away from this extension domain, allowing younger sediments to accumulate in the potential void during growth faulting (fig. 1). This downdip zone, the translation domain, is characterized by sliding along bedding-parallel ductile and brittle faults, acting as a detachment zone. If these sliding sheets of sediments reach either a sufficiently shallow slope, or an obstruction, or increased frictional resistance, their movement is impeded by buttressing, and compressive stresses shorten the sliding package in the shortening domain; thrusts and gravity-glide folds form here in the toes of the glide sheets.

Gravity sliding thus provides a mechanism of removing material from the extension domain and transporting it to the shortening domain via the translation domain, as well as accentuating subsidence in the extension domain and uplift in the shortening domain (fig. 1).

Superimposed Diapiric Strains

The zone of gravity-glide folding or thrusting moves ahead of the zone of growth faulting in time and space if the continental margin is prograding. Folding in the shortening domain results from buckling, but can be enhanced or retarded by gravity, depending on relative densities of the folded strata (fig 3).

Diapirs can be initiated in three zones.

1.   In the shortening zone far down the continental slope, gravity-glide folding triggers differential loading of synclines eventually causing buoyant diapiric rise of the anticlines when the synclines have received enough sediment. Buoyancy is previously inoperative beneath a thin cover of prodelta mud because no density inversion is present.

2. In the case of the Sigsbee Scarp, diapirs are initiated by lateral spreading of a salt nappe by mega-scale differential loading beneath the continental margin depocenter or by thermal convection.

3. Shale diapirs can be initiated in the shale half-spoon beneath the continental margin. The shale rises diapirically from the shale mounds on the landward side of each growth fault, or spreads laterally by asymmetric folds and thrusts on the continental slope.

Most of these diapiric nurseries are downdip of the extension zone of growth faulting. When the continental margin progrades over these diapiric nurseries, the diapirs may have grown large and active. Thus they can intensely deform each of the three strain domains; the Frio margin is an example. Typically doming causes local extension and uplift above domes, and local shortening adjacent to domes. This complexity makes recognition of the original strain domains difficult (fig. 4).

Key Reference: Winker (1982)

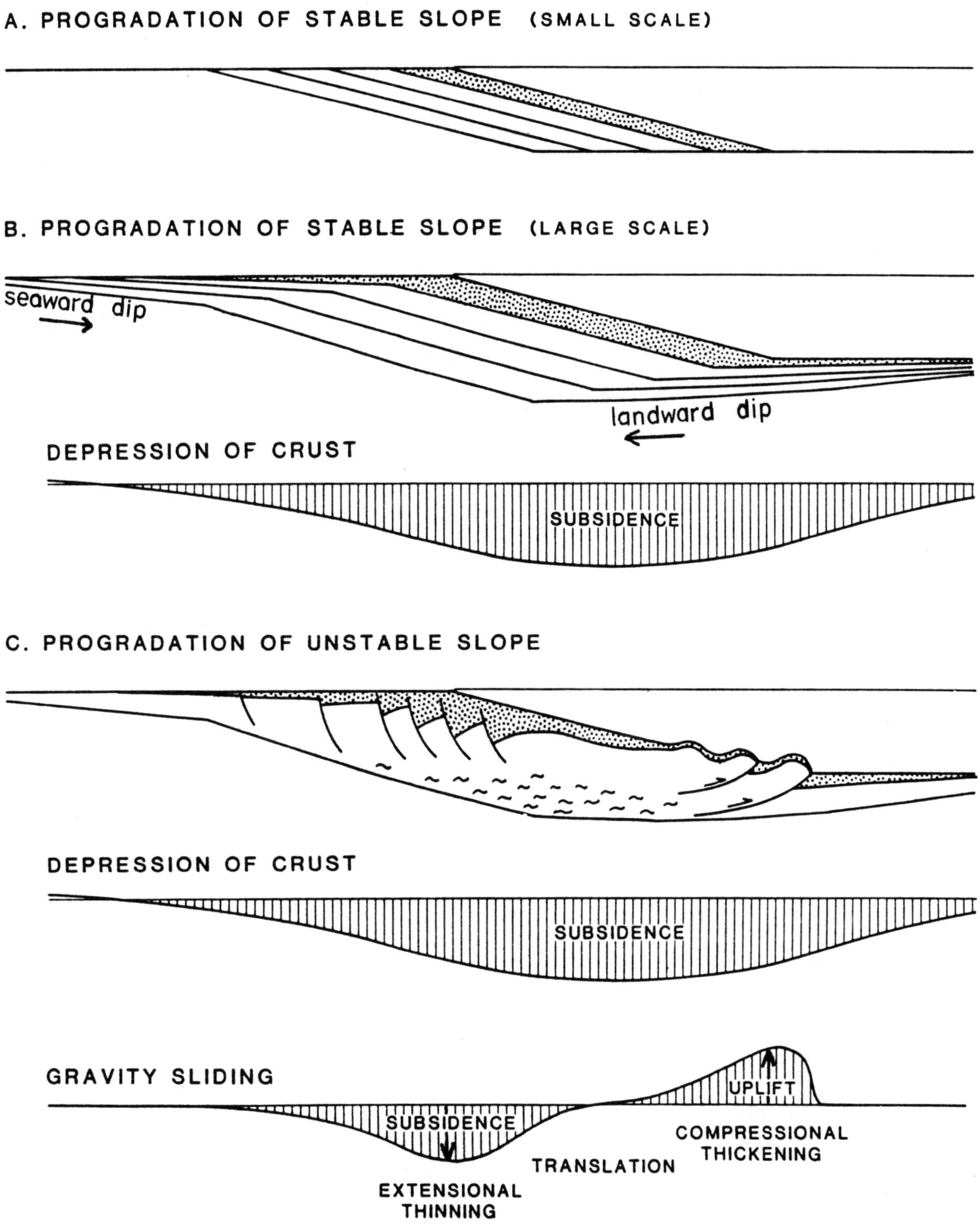

Figure 1. Dynamics of a prograding continental margin. A. No subsidence. B. Flexural subsidence of elastic lithosphere beneath the depocenter. C. Superimposed influence of gravity tectonics: extensional regime at shelf margin is characterized by growth faulting and thick sediment accumulation in fault-block sediment traps; shortening regime is characterized by gravity-glide folding or thrusting and thin sediment accumulation. (Adapted from Winker, 1982.)

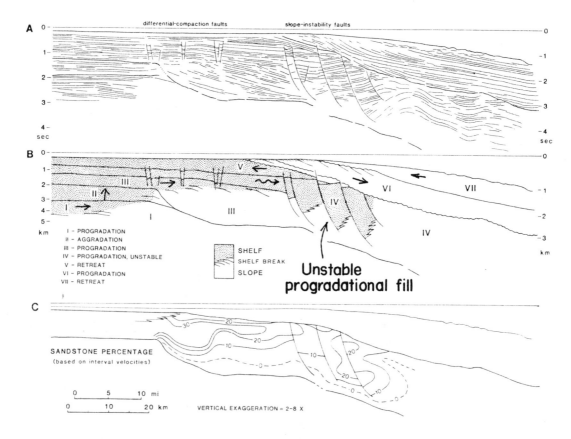

Figure 2. Interpretation of an offshore West African seismic dip section from Todd and Mitchum (1977, their fig. 9). A. Geometry; gravity-glide folding downdip of listric growth faults. B. Seismic sequences; note dramatic thickening of sequence IV in the growth-fault zone at the shelf edge. C. Concentration of sand in the growth-fault zone. (Adapted from Winker and Edwards, 1983.)

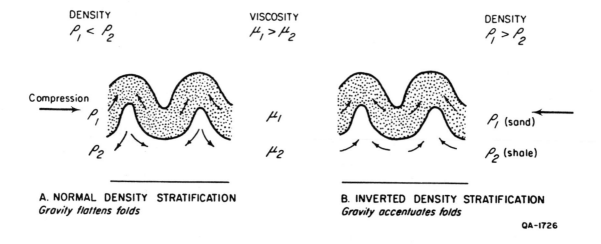

Figure 3. Effect of gravity on large (map scale) buckle folds with or without density inversion.

152

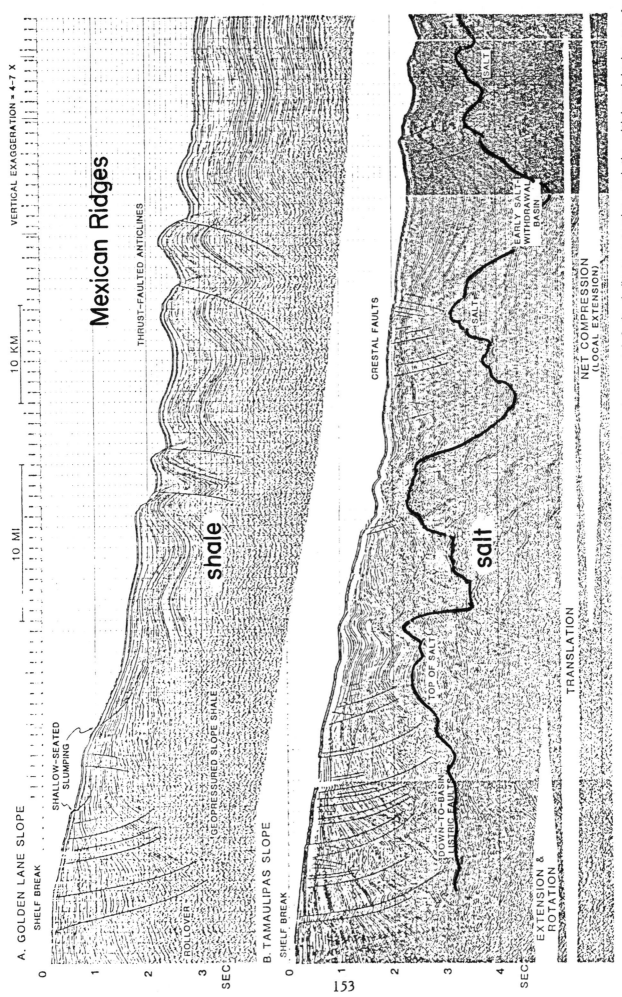

Figure 4. Recognition of continental-margin strain domains (along base of diagram) in seismic dip sections in the western Gulf of Mexico. A. Relatively simple differentiation in the Mexican Ridges. B. Salt diapirism has obscured all the strain regimes except the extensional regime at the shelf break. We do not share the view that folding above the salt necessarily indicates net lateral shortening; crestal faults indicate extension, and the withdrawal basins can form entirely by sag rather than lateral compression. In contrast the thrust faults in A are compatible with shortening above a detachment zone of nondiapiric shale. Appproximate depth conversion: 1 s = 750 to 900 m (2,500 to 3,000 ft). (Adapted from Winker, 1982.)

TERTIARY GULF COAST GROWTH FAULTING

A cross section through the growth-fault trend of South Texas is shown in figure 1. The extension domain continues onto the shelf (fig. 2). Whether these faults are due to upward propagation of older growth faults that originally marked an older shelf edge, or whether they denote weak extension caused by Gulfward creep down very gentle slopes is not known.

Growth Fault Activity and Continental Margin Migration

At the continental margin the growth fault throws and expansion indices increase markedly (fig. 2). This increase in extension and subsidence enables former shelf margins in the Gulf to be mapped, as described in unit 36.

On a smaller scale, figure 3B shows the structural evolution of the Blessing area, which lies beyond the effects of salt diapirism, as the shelf margin prograded across in Frio time. Expansion indices (= growth ratios = downthrown thickness/upthrown thickness) and rollover isopach gradients measure the activity of the growth faults. Maximum activity was coeval with the shelf margin prograding southward through the area at stages 6 and 5. In subsequent stages 4 through 1, this area lay well within the shelf, where tension was minimal, so fault activity declined greatly.

Analysis of the Brazos Ridge Growth Fault System

The Brazos Ridge is a major strike-parallel ridge of Miocene shale near the present continental margin off Texas. A master growth fault on the ridge's basinward flank (fig. 4) displaces downthrown strata laterally more than 16 km (10 mi); this lateral extension must be balanced by equivalent shortening (folds or thrusts) ESE of this section. Throw is at least 5,200 m (17,000 ft) and the expansion indices are at least 3 for the whole sequence and more than 6 for individual units. Downthrown Middle to Upper Miocene strata consist of moderately pressured (0.65 FPG) sand, silt, and shale; upthrown strata consist mainly of highly pressured (0.8 FPG) shale of Lower and Middle Miocene age capped by a reduced thickness of Upper Miocene sand and shale at depths from 2,400 m (8,000 ft) to 700 m (2,300 ft).

The master fault consists of a planar upper section which extends from the surface down to the top of the overpressured zone at a depth of about 1,500 m (5,000 ft). Below, the fault becomes listric before changing downward into a long planar ramp dipping at 9°. Rollovers in the thickened downthrown sequence are generated in the listric zone and are conveyed down the fault ramp with their rollover geometry unchanged. Heavy lines in figure 4 indicate stratigraphic intervals thickened by faulting. Thus the heavy lines show the timing of movement along each fault. For instance, the master fault has been active throughout deposition of units 1 to 12 and will probably continue to move.

The geologic history of this part of the Brazos Ridge can be reconstructed as follows. Throughout the Middle/Upper Miocene section shown (units 1 to 9) the shelf edge was close to the point where the master fault surfaces around shot point 140. Faulting carried shelf edge sands down the ramp like a conveyor belt, where they were covered by and interfingered with deep-water prodelta muds. Units 1 through 9 all show the same geometry. East of the fault they thin as they leave the rollover zone, then thicken into a broad syncline, then thin once more, in places with a counter-regional dip. The synclinal thickening is suggested to be the primary peripheral sink of a salt pillow. Rise of this hypothetical pillow caused syndepositional thinning of units 1 through 9 in the extreme east. At horizon 9 the lower part of the chevron faults in the east were briefly activated, possibly by sliding off the flanks of the pillow. The salt pillow then collapsed, possibly feeding a diapir to the east of the section.

After a brief transgression in unit 8/9 the shelf margin prograded to the extreme east of the section, initiating typical shelf-margin growth faults in the upper part of the fault chevron at the end of the Miocene. In the west many small synthetic and antithetic faults cut up through the shelf sediments of units 9 through 12 in response to continued movement on the master fault. Although no longer at the shelf margin, this fault continued to move because the Brazos Ridge forms such a gross mechanical discontinuity in the sedimentary package that this inherited structure continues to focus lateral movements as long as the continental shelf remains under tension.

Key References: Christensen (1983), Winker and Edwards (1983)

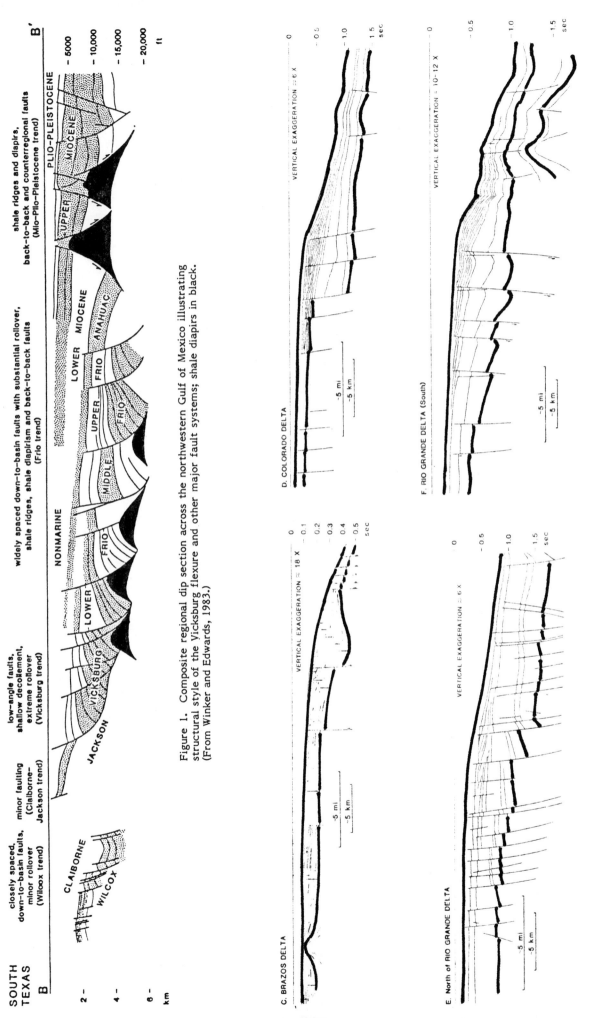

Figure 1. Composite regional dip section across the northwestern Gulf of Mexico illustrating structural style of the Vicksburg flexure and other major fault systems; shale diapirs in black. (From Winker and Edwards, 1983.)

Figure 2. Sparker dip profiles of the modern continental margin, northwestern Gulf of Mexico, showing dramatic increase in thickness of units in growth-fault zones. Approximate depth conversion: 1 s = 750 to 900 m (2,500 to 3,000 ft). (From Winker, 1982.)

157

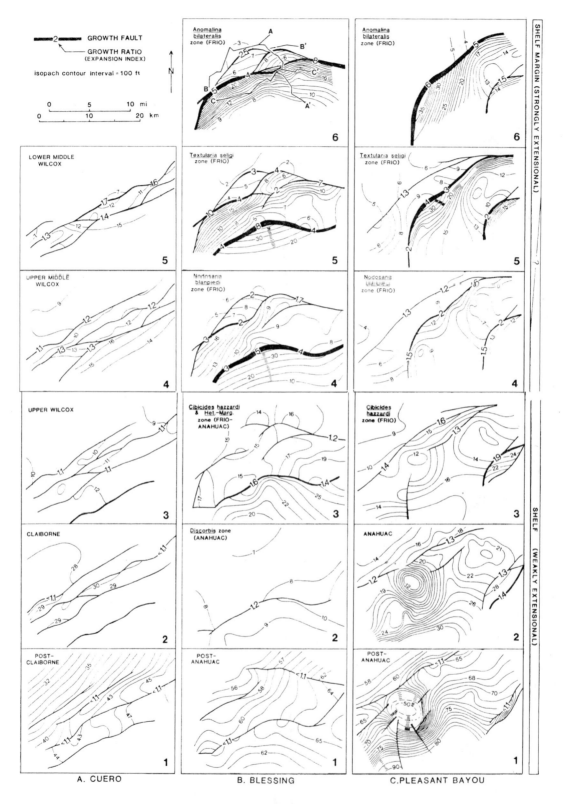

Figure 3. Structural evolution of three shelf-margin deltaic sequences in the Texas Gulf Coast illustrated by isopach maps (contour values x 100 ft). Progradation of the continental margin through each area is marked by a surge of growth-fault activity recognizable by high expansion indices and steep isopach gradients in rollover zones. C shows the effects of superimposed salt diapirism, discussed in unit 30. (From Winker and Edwards, 1983.)

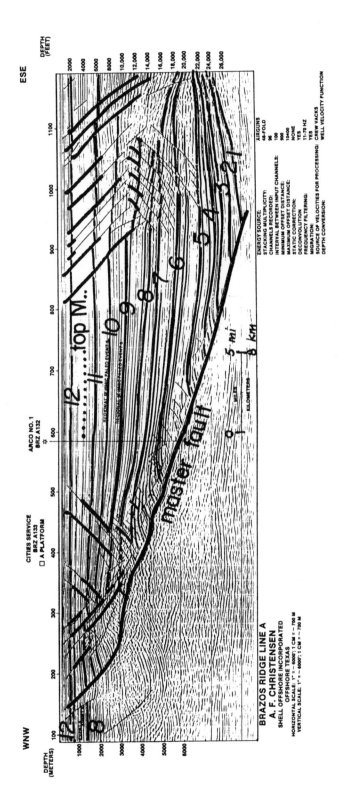

Figure 4. Seismic dip profile through the Brazos Ridge and related structures. (Adapted from Christensen, in Bally, 1983.)

159

TERTIARY GULF COAST DIAPIRISM

Distribution of Salt

Figures 1 and 2 show the distribution of salt structures in the Louann Salt underlying approximately half of the Texas-Louisiana Gulf Coast. Thinner salt is present outside these areas. The southern bulge of salt into the abyssal Gulf is due to southward creep of allochthonous salt forming the Sigsbee Scarp.

The vertically exaggerated N-S cross section in figure 3 extends from the marginal East Texas Basin across the central Gulf. The depocenter is beneath the continental margin.

The size, shape, and concentration of salt structures show a general trend from the continental slope to the abyssal plain: salt structures become older, more circular, and smaller in plan, and structurally more mature (fig. 4). They become smaller in plan because they are elongating upward in the third dimension. They become more circular because salt ridges segment into elliptical pillows which grow into subcircular diapirs by amplification of their crestal zones.

Elongated Ridges, Pillows, and Walls on the Lower Continental Slope

Highly elongated salt ridges and salt walls are best developed in the distal parts of the Mississippi and Rio Grande fans, presumably because these major depocenters exert strong loading control on salt mobilization.

The base of the Louann Salt has climbed tectonically up section by at least 5 km (17,000 ft) from its roots in the Jurassic strata virtually to the surface in Pleistocene deposits (fig. 5). Lateral movement of the wedge-shaped Sigsbee salt nappe is about 150 km (90 mi). Rise of the salt nappe has allowed immature salt structures to form near the sea bottom where the point of the wedge defines the Sigsbee Scarp. Enormous sedimentary load beneath the continental margin is squeezing salt laterally, forcing buoyant salt to move upward and sideward. The upper part of the salt nappe is convoluted by diapirism into irregular ridges (figs. 1 and 5). We would expect the wavelength of the salt ridges to increase updip as the salt source layer thickens. This is confirmed by harmonic analysis: salt ridges are initiated at a wavelength of 10 km (6 mi) on the lower slope, increasing to 27.5 km (17 mi) on the upper slope, with a mean of about 15 km (9.5 mi) (fig. 6).

How are the diapirs in the roof of the salt nappe initiated? (1) Lateral compression of the salt due to squeezing from beneath the depocenter may have caused instability in the upper surface of the salt, making it break into subdued waves with a 10-km wavelength. (2) The source layer is more than 10 km deep and is blanketed by thermally insulating sedimentary cover, which would increase the geotherm in the salt. Thus hot, expanded (less dense) salt may be rising by solid-state thermal convection; perhaps these diapirs are the near-surface expression of thermal plumes.

Pelagic deposits drape these immature salt structures. The remains of these pelagic muds may be the shale sheaths dragged up by some diapirs. Once initiated, synclinal depotroughs between the salt structures in the abyssal plain or continental rise trap mass-flow and turbidite deposits (especially the dense sands). This trapping accelerates loading of the synclines and encourages rise of salt structures by differential loading. At this stage the salt becomes diapiric, forming diapirs or walls. Once the depotroughs are deep enough to contain sufficiently compacted clastics (perhaps 1 km deep), further diapirism is driven by buoyancy as well as other forces.

As the Cenozoic continental margin progrades toward the immature salt ridges and walls, the bathymetry around the growing salt structures is transformed as the salt structures grow and structurally mature.

Large, Closely Spaced Diapiric Spines and Mounds

These structures grow beneath sediments deposited on the continental slope at moderate water depths (fig. 7). Diapiric spacing is controlled by competition for the available salt in the source layer. Piracy by the larger, more active diapirs weeds out the smaller ones that are too close to them. As water depths decrease through approach of the shelf margin, diapiric growth accelerates because of the increasing coarseness and rate of sedimentation.

Isolated Diapiric Spines

These diapirs are in their prime. They have reached equilibrium spacing and attain maximum growth rates in a zone of coarse-clastic deposition at the continental margin. Diapirs that cannot elongate and narrow sufficiently fast to keep downbuilding are smothered in this zone of extremely rapid sedimentation at the continental margin. Almost certainly many other diapirs must be deeply buried (fig. 3) and those we see near the surface, as in figure 1, are the only survivors of this weeding process.

Narrower Isolated Diapiric Spines

These diapirs have elongated markedly and are possibly close to exhausting their source layers. Here on the continental shelf and continental plain, these diapirs are rising only slowly in an environment of slow deposition. The focus of both regional subsidence and maximum deposition has passed seaward over these salt structures.

Shale Diapirs

Shale diapirs are concentrated in the Frio west of the Houston salt diapir province below the Frio continental margin (fig. 2). They formed where rates of burial were highest and where the overburden was sandiest (densest). Theoretical considerations discussed earlier in this course suggest that diapiric growth was fastest in the early stage and that further rise was choked off by the disappearance of the sand-shale density inversion by shale compaction.

Key References: Martin (1980), Stude (1978)

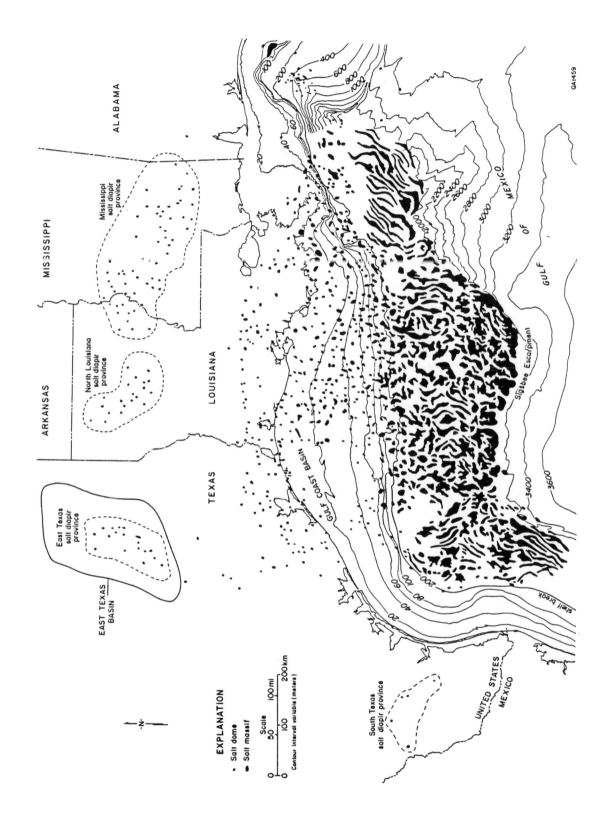

Figure 1. Salt structures in the northwestern Gulf of Mexico and adjacent interior basins.
(From Seni and Jackson, 1984.)

164

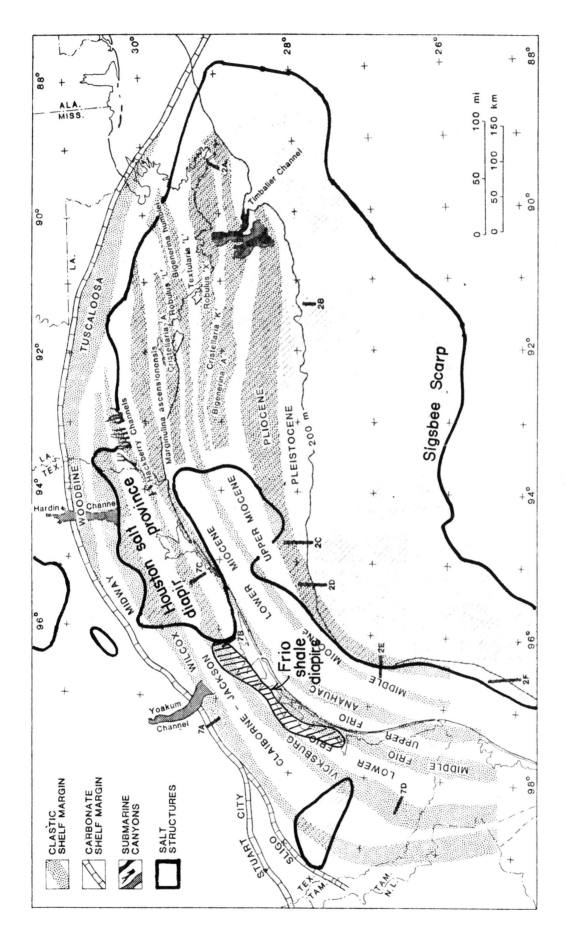

Figure 2. Distribution of salt and shale diapirs in relation to Cenozoic continental margins in the northwestern Gulf of Mexico. (Adapted from Winker, 1982.)

165

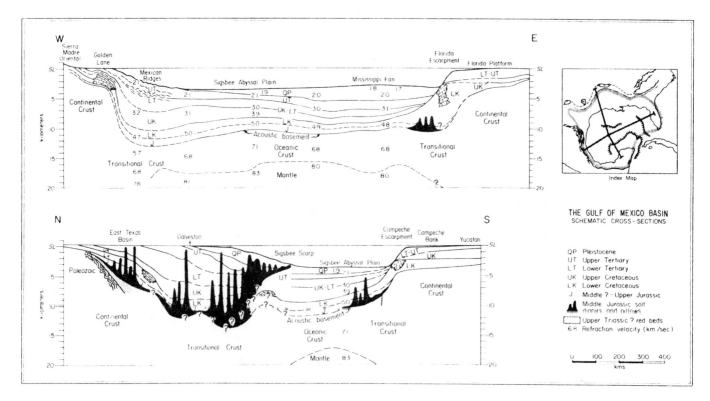

Figure 3. Schematic cross sections through the Gulf of Mexico. (From Salvador and Buffler, 1983.)

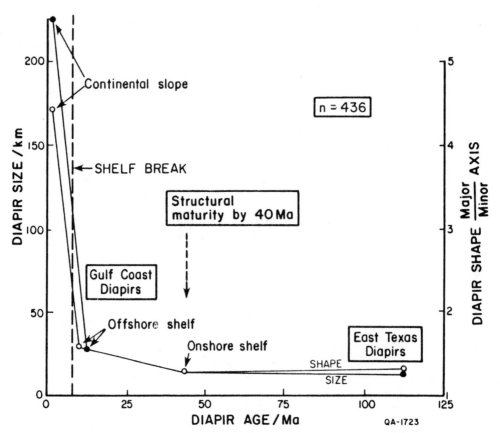

Figure 4. Salt structures become smaller and more circular with increasing age and distance inland from the continental slope, northwest Gulf of Mexico. Diapirs are evolving rapidly at the continental margin and most become structurally mature about 40 Ma after salt flow began.

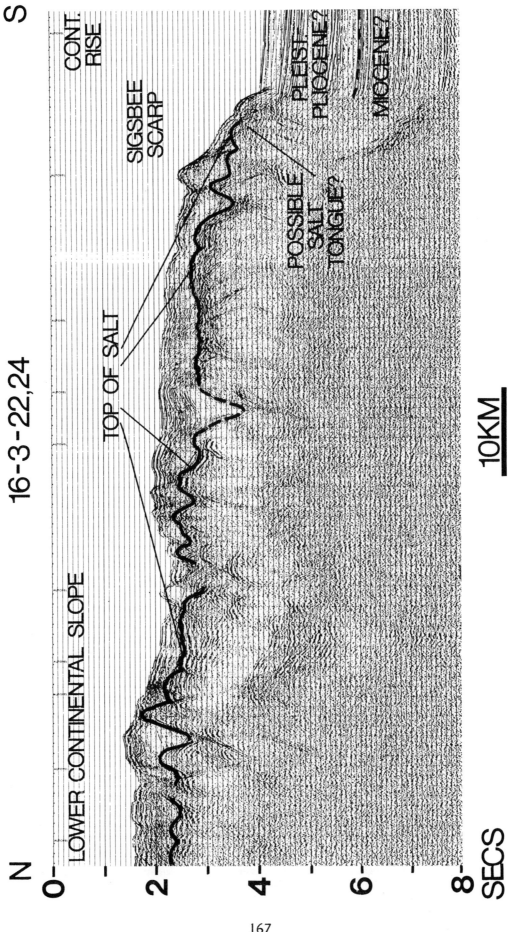

Figure 5A. Seismic dip profile across the Sigsbee Scarp, the front of a large salt nappe that has moved southward about 150 km (90 mi) into the Gulf abyssal plain. Diapirically deformed upper surface of the salt nappe. (From Buffler and others, 1978.)

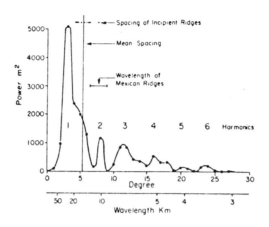

Figure 6. Power spectrum of the bathymetry shoreward from the Sigsbee Scarp. (From Lindsay, 1977.)

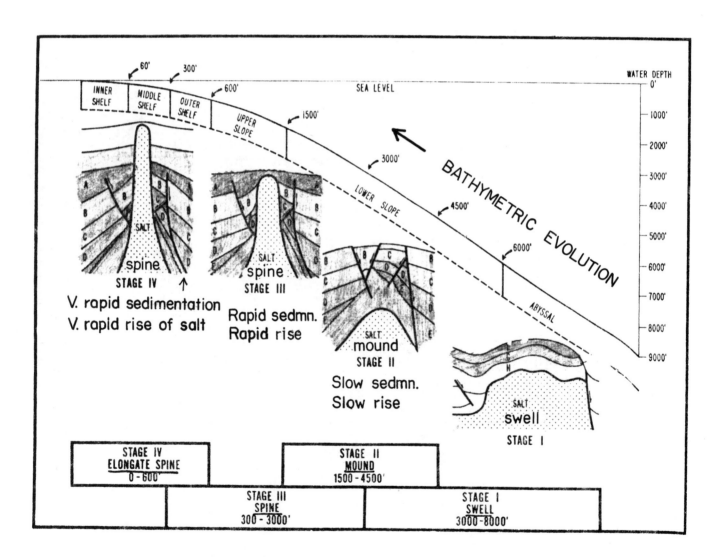

Figure 7. Relation between salt-structure evolution and bathymetric evolution induced by progradation of a continental margin. (Adapted from Stude, 1978.)

168

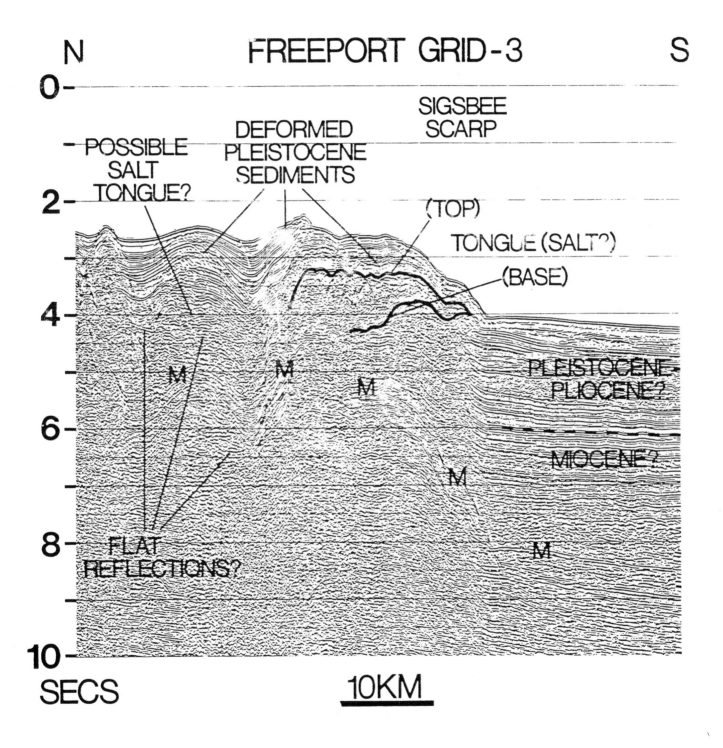

Figure 5B. Seismic dip profile across the Sigsbee Scarp, the front of a large salt nappe that has moved southward about 150 km (90 mi) into the Gulf abyssal plain. Leading edge of the salt nappe. (From Buffler and others, 1978.)

TERTIARY GULF COAST SALT DIAPIRISM AND GROWTH FAULTING INTERACTION

In the Houston salt diapir province (for locations of this and other areas described here, see unit 29) salt diapirism dominates the structural style so much that prediapiric structures have been distorted to the point of obscurity. But the growth-fault trend prominent to the southwest is still recognizable on isopach maps in the salt diapir province.

This area provides an example of the process whereby a distal zone of growing diapirs is overridden by a zone of growth faults being carried along a prograding margin. Continued rise of the diapirs deforms the growth faults. The faults may be migration paths for hydrocarbons, and fault-related traps can be distorted by salt diapirism so that hydrocarbons are redistributed or leaked in the second phase of deformation. Because of their depth and complexity, the geometry of these traps is little known and little understood, but they are likely to contain substantial reserves of geopressured gas.

Katy Area (northwest Houston diapir province)

The South Texas growth faults appear to terminate in the Orange Hill field at the top of the Wilcox Group (fig. 1). Arcuate peripheral sinks updip of the Hockley and San Felipe diapirs (which are virtually barren at upper levels) indicate salt flow from updip during their growth. Seismic data indicate that large growth faults in the Wilcox cut through the diapir trend (fig. 2). The apparent lack of withdrawal basins defined by Wilcox isopachs indicates that these diapirs had not pierced the Wilcox until the end of Wilcox time. They may have pierced pre-Wilcox levels under the influence of loading by the Wilcox deltas, but diapiric growth was minimal. By late Claiborne to Vicksburg time, the diapirs matured to form overhangs and overturned strata in their zones of contact strain.

Pleasant Bayou Area (southwest Houston diapir province)

Structures like the Danbury Dome deform regional Frio growth faults in Anahuac time, as shown in figure 3C of unit 28. These faults were extremely vigorous (growth ratios up to 6) as the shelf margin passed through this area. The dome rose where the largest growth fault curved sharply. Its rise disrupted the growth fault, which was already waning, and no further faulting took place. The dome is not a piercement structure at this level.

171

Port Arthur Area

The structure map on top of the Frio Formation (fig. 3) shows regional growth faults and a line of three salt diapirs: Spindletop Dome, a shallow, oil-rich piercement dome, and Port Neches and Orange Domes, which have not pierced the Frio.   In early-to-middle Frio time, broad salt pillows acted as reservoirs for later diapiric growth (fig. 4).   Spindletop probably drew salt from the two flanking pillows along the E-W salt trend (fig. 4), aided by strong subsidence downdip of a major growth fault to the north (fig. 5).   By Anahuac time the other two domes had concentrated the salt from an originally broad pillow (fig. 4) into two smaller, taller, salt pillows (fig. 5).

This area is notable for the comparative mildness of growth faulting during the Frio (figs. 4 and 5), probably because the area was dominated by mud and silt deposited comparatively slowly, away from the principal Frio sandy deltas.

Key Reference: Ewing (1983)

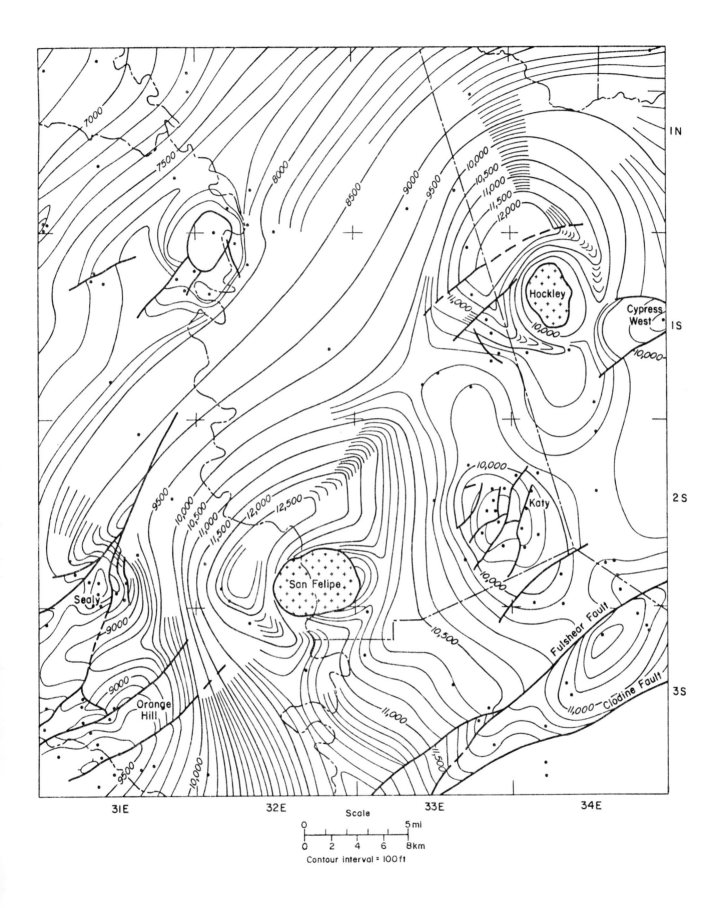

Figure 1. Structure contours on top of the Wilcox Group, Katy area, Texas. (From Ewing, 1983.)

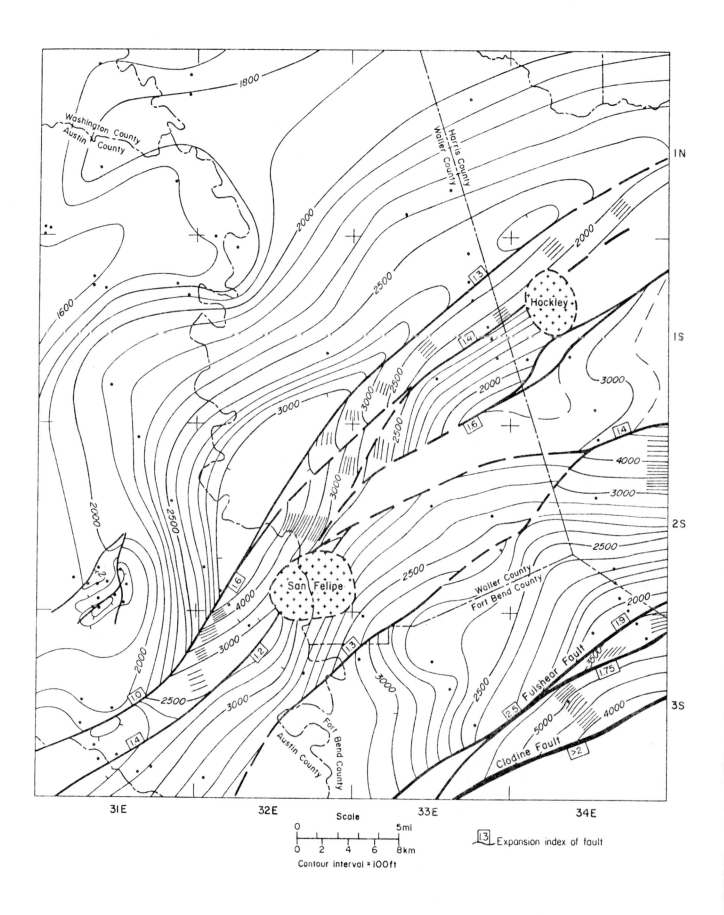

Figure 2. Isopachs for the middle-upper Wilcox Group, Katy area, Texas. (From Ewing, 1983.)

174

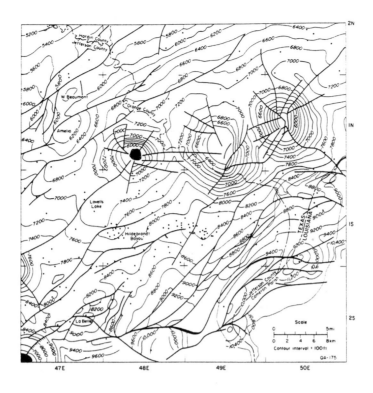

Figure 3. Structure contours on top of the Frio Formation, Port Arthur area, Texas. (Adapted from Ewing, 1983.)

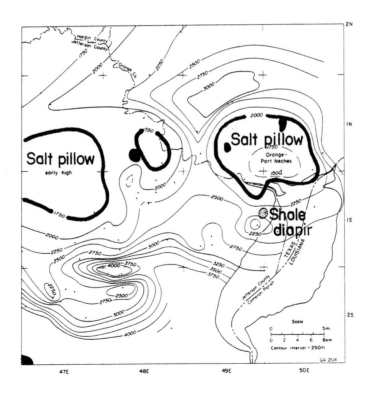

Figure 4. Isopachs for the lower-middle Frio Formation, Port Arthur area, Texas; minor faults omitted. (Adapted from Ewing, 1983.)

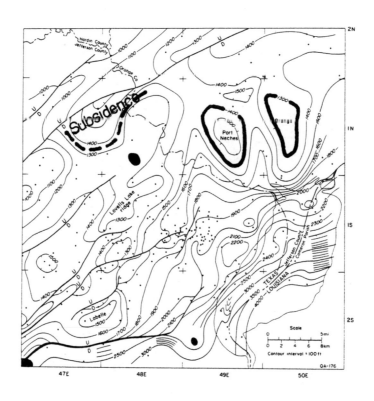

Figure 5. Isopachs for the upper Frio and Anahuac Formations, Port Arthur area, Texas; minor faults omitted. (Adapted from Ewing, 1983.)

TERTIARY GULF COAST GRAVITY-GLIDE FOLDING AND FAULTING: THE MEXICAN
RIDGES

The Mexican Ridges underlie almost all of the western Gulf of Mexico south of the Rio
Grande.   The southern half of the ridges has been studied in the most detail, resulting in
subdivision into two zones--4a and 4b (fig. 1).   The reader is referred to unit 27 for a discussion
of the tectonic setting of fold and thrust zones on divergent continental margins.

Zone 4a: Thrust Sheets and Gravity Gliding

Zone 4a is the simpler of the two.   Strike-parallel, symmetric folds have an average
wavelength of 10 to 12 km (6.5 to 7.5 mi) and structural and bathymetric relief of several
hundred meters (fig. 2).   The folds tightened seaward serially in the Pleistocene by a minimum
of 1 percent shortening of a 3-km (2-mi) thick allochthonous sheet gliding over a detachment
zone of thick, Upper Cretaceous-Lower Tertiary pelagic shale.   These gravity-glide folds are
shale-cored and produced a minimum lateral displacement of 1 to 2 km (0.6 to 1.2 mi).   A deep-
rooted diapiric structure has caused updip folds to pile up against this buttress (fig. 3); its stress
shadow has prevented folds forming directly downdip.

Three folded thrust sheets form the entire allochthon of zone 4a.   The timing of their
movement can be deduced from relations with the sediment veneer above the folds (fig. 4).
Thrust sheet 1 formed near the top of the overpressured shales.   The extension domain is
marked by growth faults at the present shelf break.   The shortening domain is represented by
asymmetric folds with forelimb thrusts.   Onlap of undeformed Pleistocene strata against the
anticlines indicates further sedimentation after this folding.   Thrust sheet 2 formed after the
compressive stresses had migrated in front of the first thrust sheet.   Sheet 2 and its veneer
were shortened by folding and internal thrusting in the same way as sheet 1, except that an
early anticline formed over the ramp of the thrust, where more competent sands give way to
shale seaward.   Thrust sheet 3 formed downdip of the others.   It is more gently folded and is
probably still contracting.

Zone 4b: Growth Faults and Gravity-Glide Folding

Zone 4a is separated from zone 4b by a linear, dip-oriented discontinuity, which is
probably a tear fault (fig. 1).   Zone 4b is the more complex (fig. 5).   Deformation started in the

middle Miocene and resulted in considerable structural relief that rarely has bathymetric expression. Asymmetric, strike-parallel folds formed by gravity gliding and show strong seaward vergence. Anticlinal cores contain shale that became diapiric in the late Miocene as a result of turbidite loading of synclines (fig. 6). Loading stopped gravity gliding and eventually caused the steep, seaward facing fold limb of each anticline to fail in the form of a seaward-dipping growth fault. Rotation and subsidence of the downthrown side of each normal fault accentuated sedimentary loading in the synclines, which encouraged further subsidence on the downthrown side of each fault. Shale rose diapirically on the upthrown side of each fault, further amplifying the original anticlines. The weak bathymetric expression of these anticlines and of their associated growth faults suggests that the shale is no longer diapiric after an increase in density or viscosity through compaction.

Key References: Buffler and others (1979), Pew (1982)

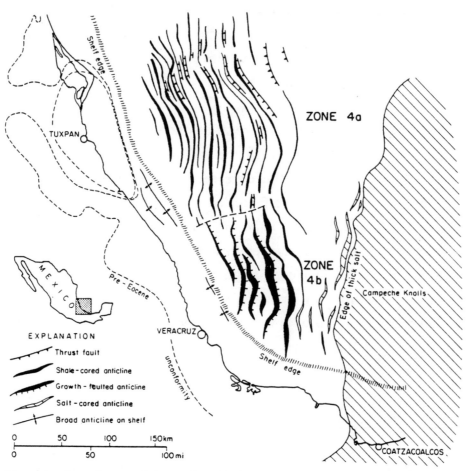

Figure 1. Tectonic map of the southern Mexican Ridges, western Gulf of Mexico. (Adapted from Pew, 1982.)

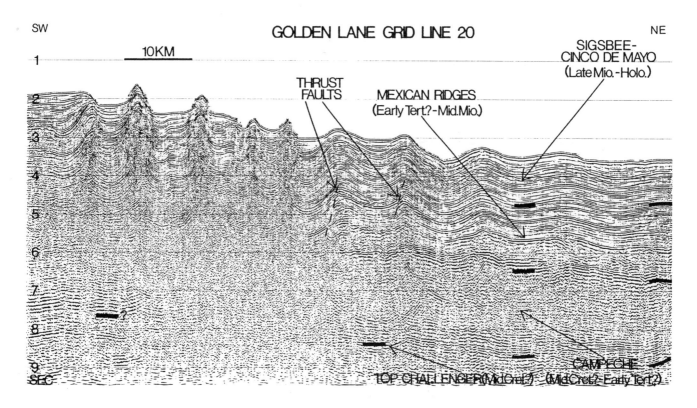

SIGSBEE-
CINCO DE MAYO
(Late Mio.-Holo.)

THRUST
FAULTS

MEXICAN RIDGES
(Early Tert.?-Mid.Mio.)

CAMPECHE

TOP CHALLENGER (Mid.Cret.?) - (Mid.Cret.?-Early Tert.?)

Figure 2. Isopachs for the middle-upper Wilcox Group, Katy area, Texas. (From Ewing, 1983.)

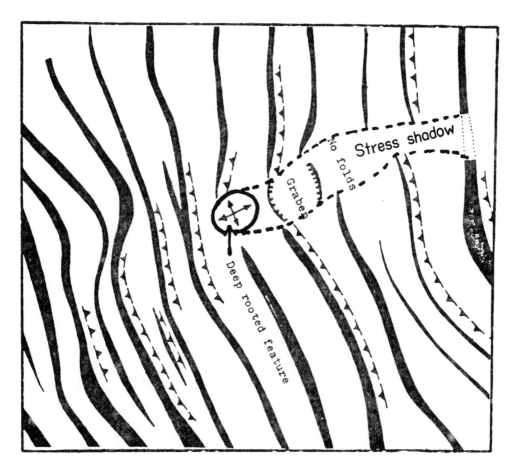

Figure 3. Structure contours on top of the Frio Formation, Port Arthur area, Texas. (Adapted from Ewing, 1983.)

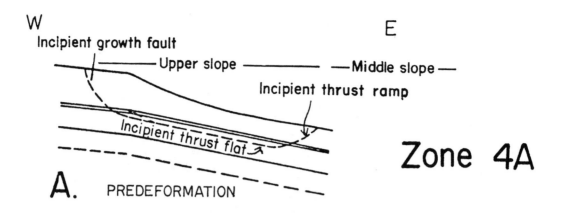

W   E

Incipient growth fault

Upper slope —————— Middle slope —

Incipient thrust ramp

Incipient thrust flat

Zone 4A

A.   PREDEFORMATION

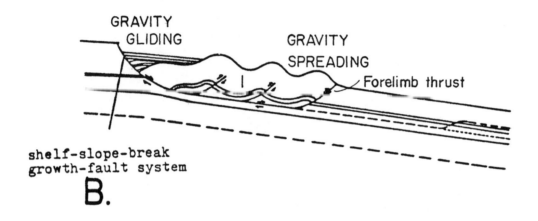

GRAVITY GLIDING   GRAVITY SPREADING

Forelimb thrust

1

shelf-slope-break growth-fault system

B.

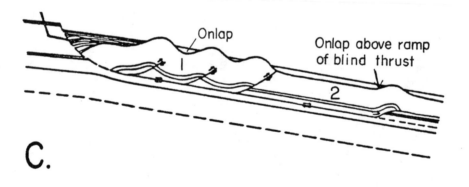

Onlap

Onlap above ramp of blind thrust

2

C.

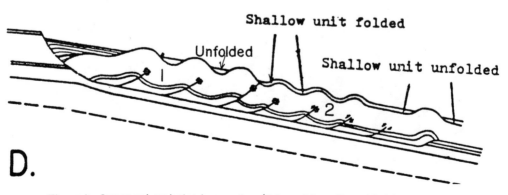

Shallow unit folded

Unfolded

Shallow unit unfolded

1

2

D.

Figure 4. Structural evolution in zone 4a. (Adapted from Pew, 1983.)

180

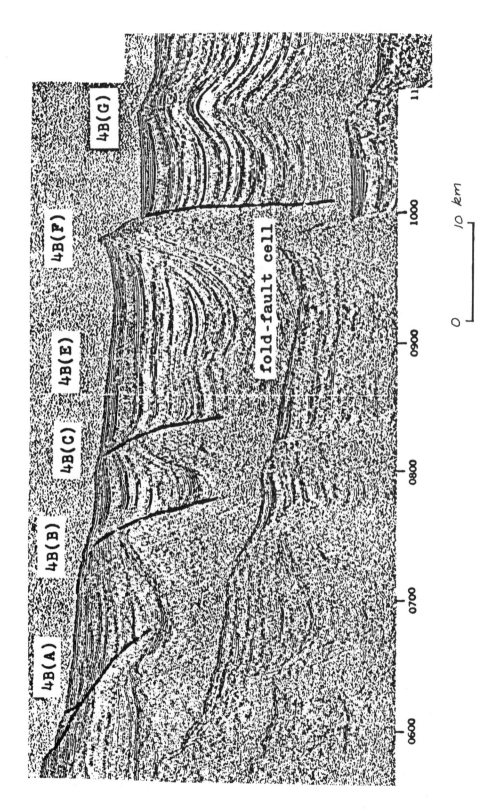

Figure 5. Seismic dip profile through zone 4b showing growth faults and shale mounds. (From Pew, 1983.)

181

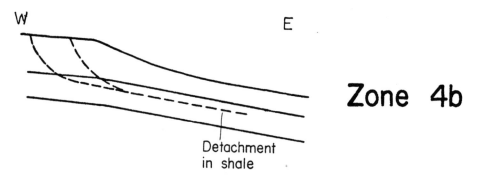

Zone 4b

Detachment
in shale

A. FORMATION OF DETACHMENT

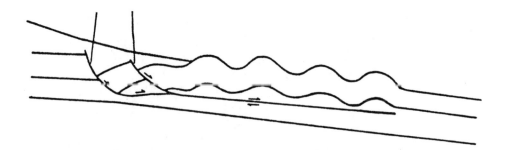

B. EARLY GRAVITY SLIDING

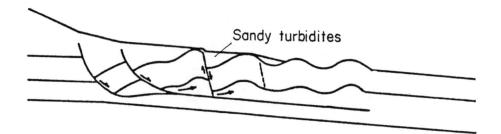

Sandy turbidites

C.
END OF GRAVITY SLIDING, EARLY LOADING & DOWNBUILDING

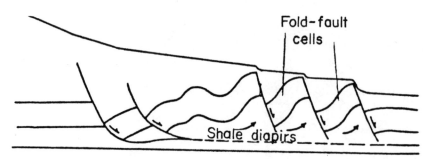

Fold-fault
cells

Shale diapirs

D.
LOADING, DOWNBUILDING, GROWTH FAULTING . & DIAPIRIC
FOLD GROWTH

Figure 6. Structural evolution in zone 4b. (Adapted from Pew, 1983.)

GULF COAST TERTIARY DEPOSITIONAL UNITS

The overall depositional history of the Northwest Shelf of the Gulf of Mexico reflects multiple cycles of clastic influx (depositional episodes) punctuated by thin, occasionally regional transgressions. Regional stratigraphic analysis of the major clastic episodes shows two types (fig.1).

1. Major episodes rapidly prograded to the existing continental margin and resulted in widespread construction of new continental platform. Slope offlap resulted.

2. Minor episodes of sediment input prograded the shoreline only partially across the extant submerged continental platform. The shelf margin remained a zone of slow pelagic sedimentation, sediment bypass, or possibly erosion. The slope may have been blanketed by spillover of distal prodelta and shelf suspended sediments or low-density turbidity currents. Little opportunity for deposition of sand at the shelf margin or onto the slope existed.

A tentative classification of Tertiary depositional episodes is shown in figure 2. For comparison, the episodes, known subregional inner coastal plain unconformities, regional transgressive marine shales, and onlap or gorge fills (reflecting submarine erosion and deep marine deposition landward of the general continental margin) are plotted beside the Cenozoic cycle chart. It appears that in a depositionally active trailing-edge margin, such as the Northern Gulf, sediment supply or other controls supersede eustatic cycles in determining depositional history.

Principal sediment input shifted among three axes -- the Rio Grande, Houston, and Mississippi embayments (fig. 3). Major delta systems prograded a sand-rich continental margin at the basinward margin of these depoaxes.

Major offlap episodes mobilize underlying thick salt. Such mobilizing units are responsible for generating salt structures. However, because active mobilization occurs beneath the advancing slope wedge, resedimented sands are deposited between rather than over structures. Later depositional episodes may inherit the salt structures mobilized by earlier episodes and deposit deltaic and shore-zone reservoirs across the structures. Thus, on a per-reservoir-volume basis, the Frio and Yegua Formations (embanking units) far exceed the Wilcox (a mobilizer) in productivity in the Houston Embayment.

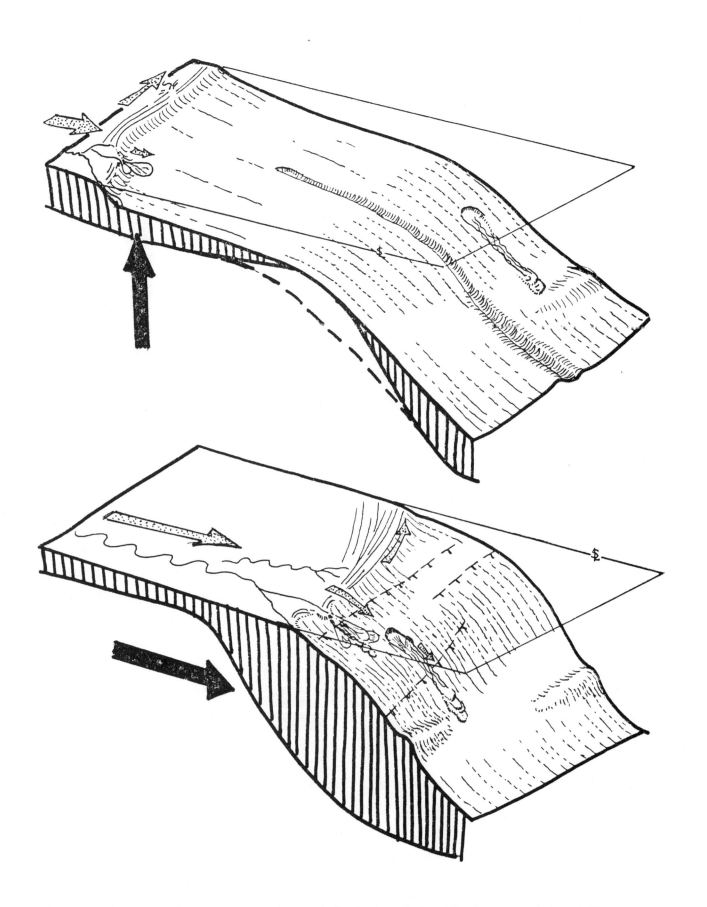

Figure 1. Contrasting morphology and distribution of depositional systems during shelf aggradation and slope offlap depositional episodes.

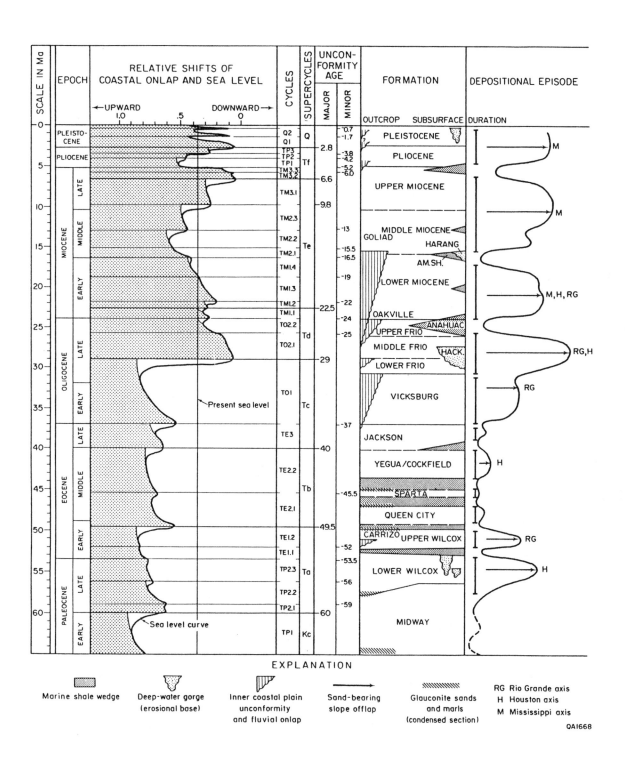

Figure 2. Temporal sequence of Gulf Coast Cenozoic depositional episodes, subaerial and submarine unconformities, and widespread marine shales.

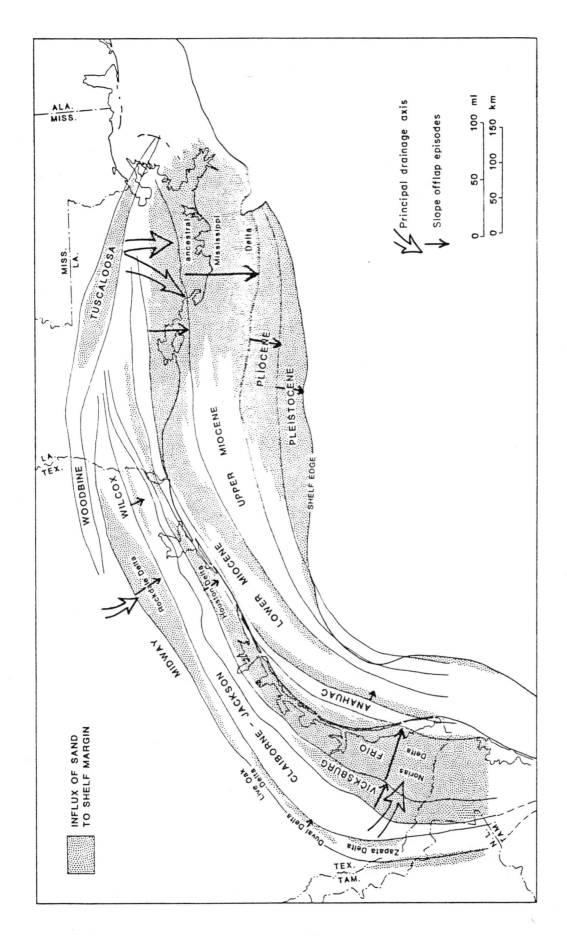

Figure 3. Shifting position of the major Cenozoic, sand-rich, deltaic continental margin. From Winker (1982).

FRIO FORMATION -- EMBANKING SLOPE OFFLAP EPISODE

The Frio Formation of the Texas Coastal Plain consists of two major delta systems and an interdeltaic wave-dominated shore-zone system, all fronted by the shelf margin and continental slope (fig. 1). Depositional style of the Frio is amazingly similar to that of the Quaternary fig. 2). However, the major river entered along the Rio Grande rather than the Mississippi axis.

Three paleomargin sequences (fig. 1) illustrate depositional and structural features typical of unstable progradational margins.

1. The Pleasant/Chocolate Bayou area lies within a salt-withdrawal basin filled with lower Frio deltaic and upper slope sediments (fig. 3). Structure includes both major bounding growth faults and deep salt domes (fig. 4). (Structural evolution was discussed in unit 28). Regional down-to-basin faults produce local roll-over (fig. 5), and more than five-fold expansion of the stratigraphic section (fig. 6).

Despite the rapid structural growth typical of a continental margin setting, lithologic features, sand-body geometry, electric-log patterns, and paleobathymetry all show that the lower Frio (Anomalina bilateralis zone) reservoirs were deposited as delta fringe sands of the Houston delta system (tables 1 and 2).

2. Together, data from the Blessing and Corpus Christi Channel areas illustrate paleomargin facies and structural development along the barrier/strandplain system. Here, underlying thick salt is absent, and structural styles are a product of shale tectonics.

Beyond the underlying Vicksburg margin, the distal shoreface and shelf sands of the Greta/Carancahua system show great expansion across a succession of growth faults (fig. 7). Both the Blessing and Corpus Christi Channel areas are bounded by such fault zones (fig. 8).

Depositional facies of the Blessing area are summarized in tables 3 and 4. Figures 9 and 10 illustrate sand-body geometry and log character of distal-barrier shoreface and associated sands of the storm-punctuated shelf. Shelf margin and upper slope deposits are faulted below well penetration. However, the cross section and map show that sand transport to the continental margin is limited.

Key References: Galloway et al. (1982 a and b); Winker et al. (1983)

Table 1. Characteristics of operational stratigraphic units used in studies of the Pleasant Bayou area. From Winker et al. (1983).

Unit numbers for isopach maps and structural sections	Log-correlation marker	Benthic foraminiferal zone	Stratigraphic unit	Lithology	Relative velocity	Seismic signature	Overall transgressive/regressive character
1			Undifferentiated Miocene and younger	High sandstone; some "hard streaks" (high-carbonate beds) in lower part of section	High; approximately linear increase with depth	Variable amplitude and low continuity high in section, grading down into high amplitude and high continuity	Regressive
	TA						
2		*Discorbis* and *Het.-Marg.*	Anahuac	Shale (poorly indurated)	Low	Virtually transparent (coherent reflectors may be multiples)	Transgressive
	T2						
3		*Cibicides hazzardi*	Upper Frio	Interbedded sandstone and shale; sandstone percentage generally decreases up-section	High; general increase with depth. No apparent relationship with geopressure distribution	Moderate amplitude and continuity. On many sections, obscured by multiples from Miocene section	Transgressive (stacked regressive cycles)
	T3						
4		*Nodosaria blanpiedi*	Middle Frio				
	T4						
5		*Textularia mississippiensis*	Lower Frio				
	T5						
6		*Anomalina bilateralis*	Vicksburg	Shale	Low	Low amplitude; often obscured by multiples	Regressive
7		*Textularia warreni*					

Environmental interpretation	Log characteristics	Geographic distribution
Fluvial/deltaic plain and channels	High sandstone; blocky patterns; laterally discontinuous; total section thin	Updip of main growth fault
Distributary channel and channel-mouth bar	Blocky to upward-fining pattern. Thick sandstones with few shale breaks. Laterally continuous with delta-front facies	Downdip of main growth fault; patchy distribution
Delta front	Thin to thick, upward-coarsening sandstones. Laterally continuous	Widely distributed downdip of main growth fault
Delta front in rapid-expansion zone (very high subsidence rate)	Many thin sands interbedded with shale; difficult to correlate; very thick section	Near main growth fault on downthrown side

Table 2. Electric-log facies in the lower Frio <u>Anomalina bilateralis</u> zone, Pleasant Bayou area. From Winker et al. (1983).

Unit numbers for isopach maps and structural sections	Log-correlation marker	Benthic foraminiferal zone	Stratigraphic unit	Lithology	Relative velocity	Seismic signature	Overall transgressive/regressive character
1			Undifferentiated Miocene and younger	Thin sandstones, in shale; net thickness increases downdip	Fairly high; approximately linear increase with depth	Moderate to high amplitude and continuity (highest at base of unit)	Regressive
	— B1						
2		Discorbis	Anahuac	Shale	Low to moderate, no increase with depth	Virtually transparent	Transgressive
	— B2						
3		Het.-Marg. and Cibicides hazzardi	Upper Frio	Shale; thin sandstones near base	Moderate, increasing with depth	Transparent at top, high amplitude at base	
	— B3						
4		Nodosaria blanpiedi	Middle Frio	Sandstone and shale, mostly thin sandstone	Moderate to high, increasing with depth	Moderate to high amplitude and continuity	Transgressive (stacked regressive cycles)
	— B4						
5		Textularia seligi/Text. mississippiensis	Lower Frio	Sandstone and shale, moderate amounts of sandstone	Fairly high, slight increase with depth	Moderate to high amplitude and continuity	
	— B5						
6		Anomalina bilateralis		Sandstone and shale, moderate amounts of sandstone	Fairly high, slight increase with depth	Low to moderate amplitude and continuity	Regressive
	— B6						
7		Textularia warreni	Vicksburg	Shale	Probably low	Generally low to moderate amplitude	

Table 3. Characteristics of operational stratigraphic units used in studies of the Blessing area. From Winker et al. (1983).

Environmental interpretation		Log characteristics	Geographic distribution
Fluvial/ deltaic plain and channels	low subsidence rate	Thin, laterally discontinuous sandstones	Updip of growth faults
Coastal barrier/ strand-plain, possibly some channels	moderate subsidence rate	Thin, laterally persistent sandstones, mostly upward coarsening	Fault block north of Blessing; southern part of Blessing fault block
	high subsidence rate	Thick, laterally persistent sandstones, few shale breaks, log pattern variable	Center of Blessing fault block
	very high subsidence rate	Numerous thin sandstones, abundant shale breaks, laterally continuous with thick sandstone facies	Northern part of Blessing fault block, near major growth fault

Table 4. Electric-log facies in the lower Frio Anomalina bilateralis zone, Blessing area. From Winker et al. (1983).

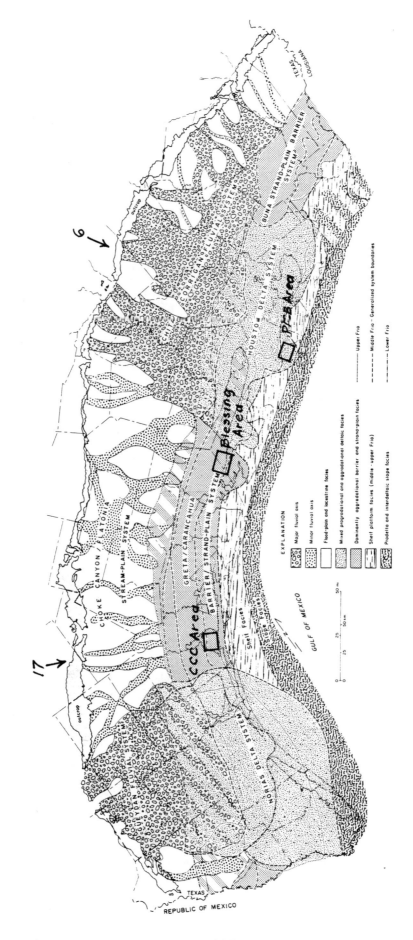

Figure 1. Depositional systems of the Frio Formation. From Galloway et al. (1982a).

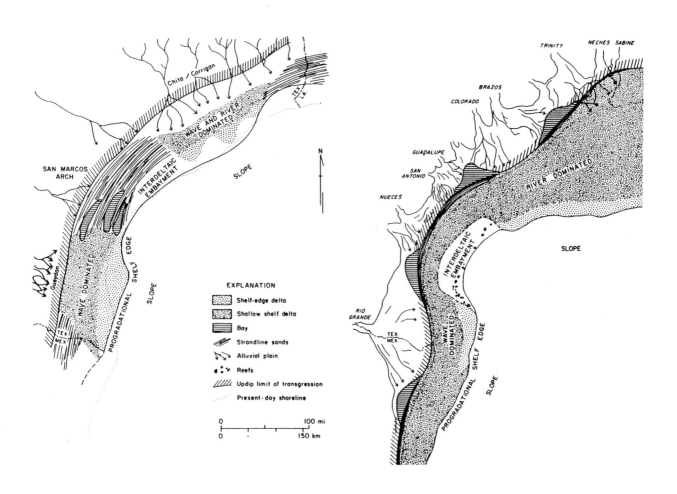

Figure 2. Comparative Frio and late Pleistocene paleogeography of the Texas coastal plain and continental margin. From Galloway et al. (1982b).

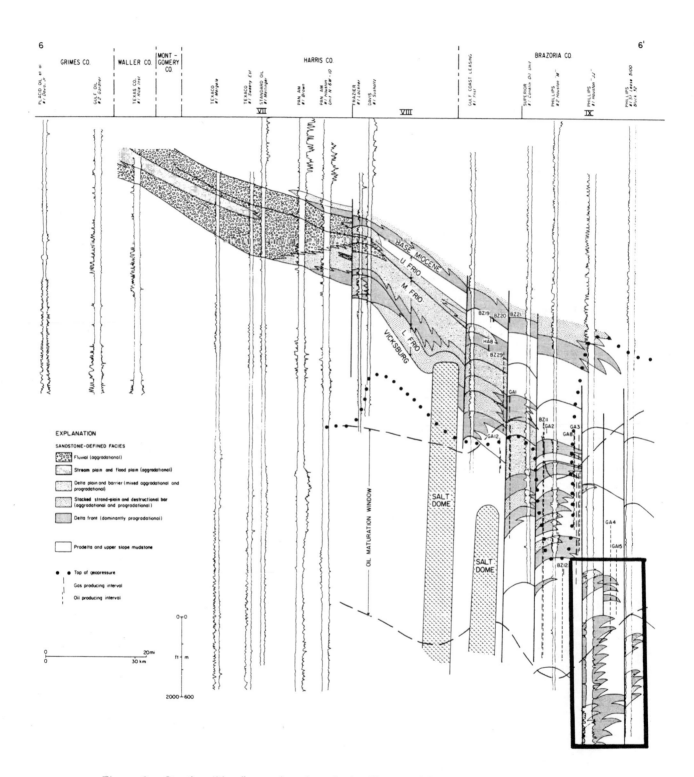

Figure 3. Stratigraphic dip section through the Houston delta system. The stratigraphic position of paleomargin deposits of the Pleasant Bayou area is indicated by the box. From Galloway et al. (1982b).

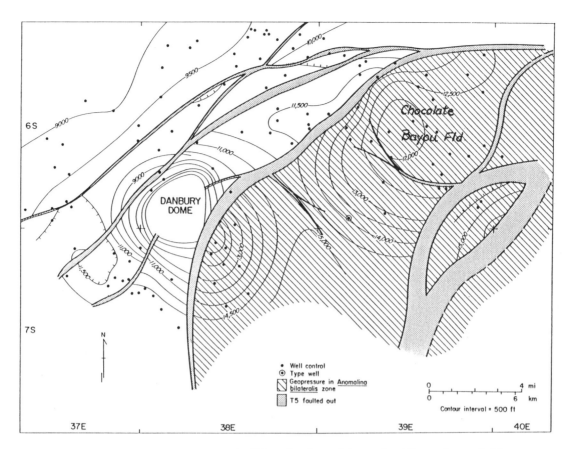

Figure 4. Structure-contour map on the T5 marker (top of <u>Anomalina bilateralis</u> zone), Pleasant and Chocolate Bayou study area. From Winker et al. (1983).

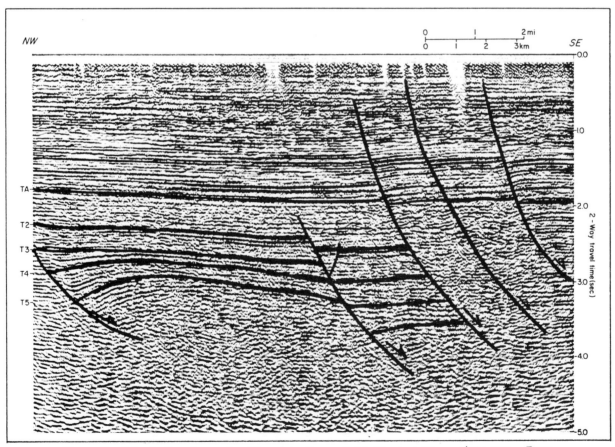

Figure 5. Seismic line traversing Chocolate Bayou field, Pleasant Bayou study area. From Winker et al. (1983).

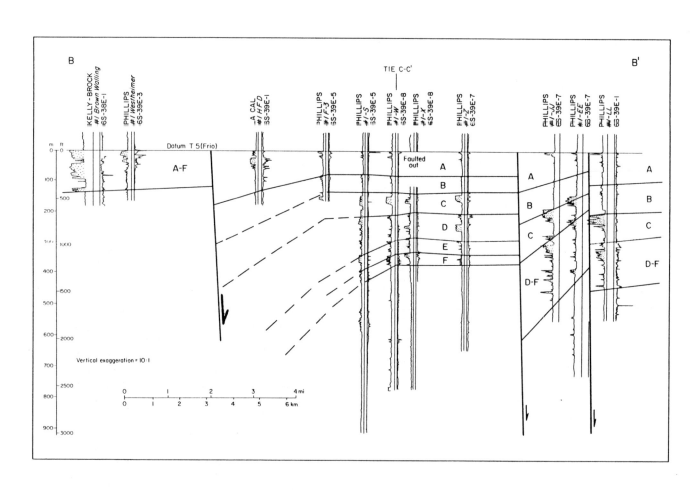

Figure 6. Stratigraphic dip section of lower Frio (<u>Anomalina bilateralis</u> zone), illustrating changes in sand-body thickness and log character due to the extreme vertical expansion across the master growth fault. From Winker et al. (1983).

196

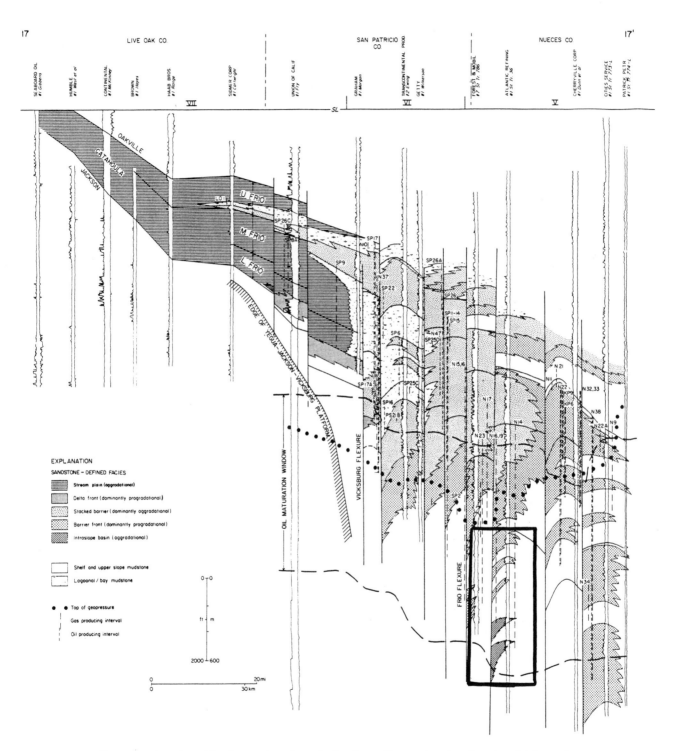

Figure 7. Stratigraphic dip section through the Greta-Carancahua barrier-strandplain system. Stacked barrier and associated shelf facies are particularly well developed in this portion of the Frio paleomargin. The box outlines the stratigraphic position of sandstones mapped in the Corpus Christi Channel area. From Galloway et al. (1982b).

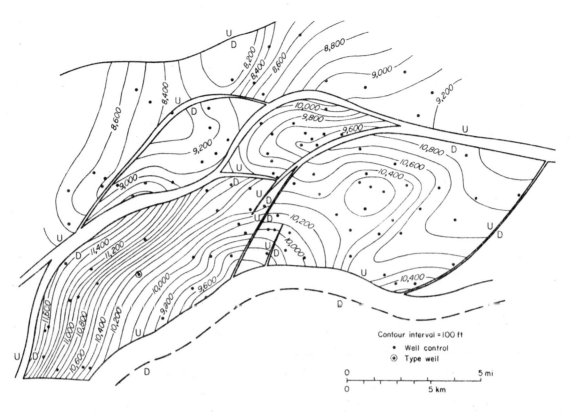

Figure 8. Structure-contour map on the B5 correlation marker (top of _Anomalina bilateralis_ zone), Blessing study area. From Winker et al. (1983).

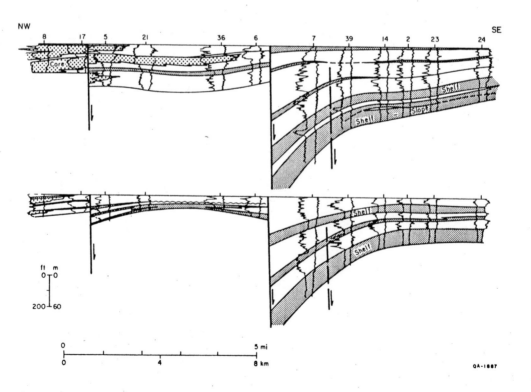

Figure 9. Cross sections of lower Frio distal-barrier shoreface and shelf sand bodies, Corpus Christi Channel area.

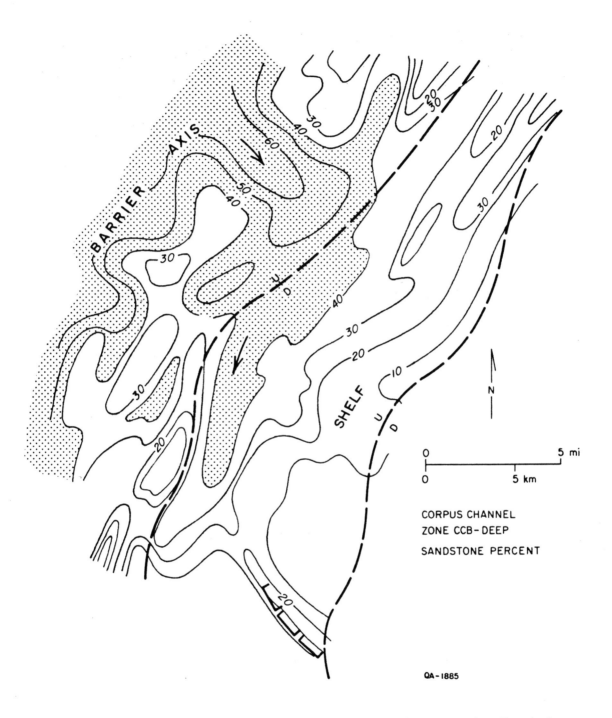

Figure 10. Distribution of a fore-barrier shelf sandstone. Contour trends reflect both structural and depositional process influences. The contemporary shelf margin probably lay downdip of the southeastern growth fault, below the depth of well data.

HACKBERRY EMBAYMENT -- SLOPE ONLAP WEDGE

A deep-water marine wedge, the Hackberry slope system, extends from the eastern flank of the Houston delta system along strike into western Louisiana. The Hackberry consists of a lower sequence containing sands and an upper sequence dominated by shale (fig. 1). The unit is floored by a widespread unconformity that truncates middle, and locally lower, Frio shelf and paralic deposits. The unconformity displays considerable channeling and local truncation of underlying Frio section exceeds 1,000 ft (600 m). Hackberry sandstones occur mainly within the incised canyons.

Fill of the canyons contains a well-described bathyal fauna. Sands display sedimentary features typical of slope systems.

Upper Frio deltaic and shore-zone systems prograde across the Hackberry wedge (fig. 1). Actual time represented by slope canyon cut and fill is thus limited to about 2 to 3 million years.

Though not obviously related to a regional transgressive shale, such as the middle Wilcox Yoakum gorge, the Hackberry occupies a common site of submarine canyon cutting -- the flank of a major delta system. Similarly located submarine canyon fills are found on the flanks of the Tertiary Niger delta system.

The upper slope canyon system is quite complex (fig. 2). Canyon fill onlaps the underlying unconformity, showing that the gorges were scoured and then backfilled, much like their Quaternary counterparts flanking the Mississippi delta system. Gorge-fill sand geometry is controlled by the canyon topography and syndepositional structural development (fig. 3). Sand-rich pockets occur within the upper slope. Although log patterns indicate both channel and suprafan deposition, the limited size and life span, onlap depositional architecture, and complex slope bathymetry of the Hackberry slope system bring into question the assumption that classical sand-rich fans were deposited on the lower Hackberry slope.

Key References: Ewing and Reed (in press); Paine (1971)

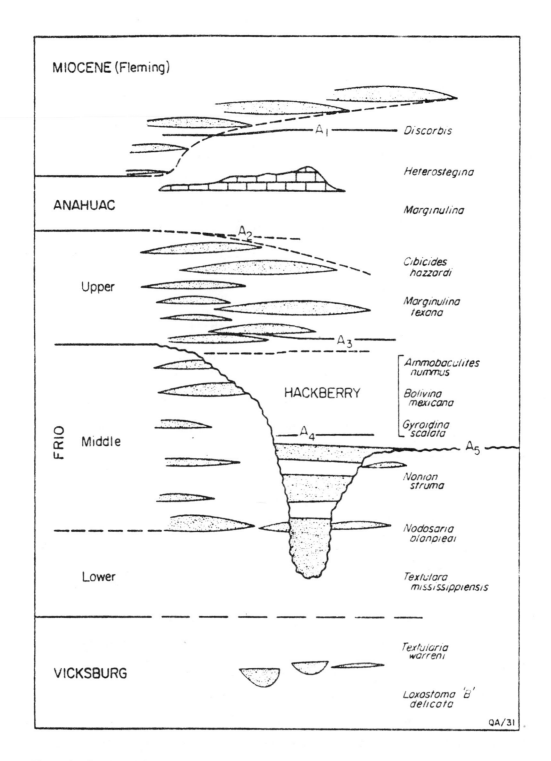

Figure 1. Stratigraphic setting of the Hackberry onlap slope system. From Ewing and Reed (in press).

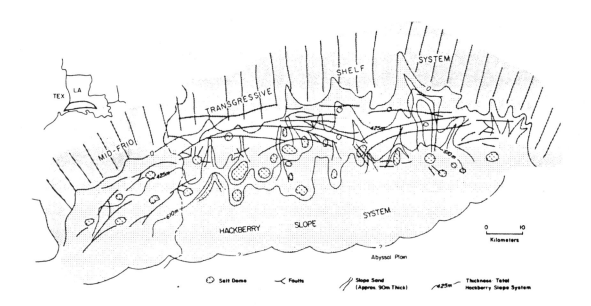

Figure 2. Depositional overview of the Hackberry slope system. From Brown and Fisher (1977).

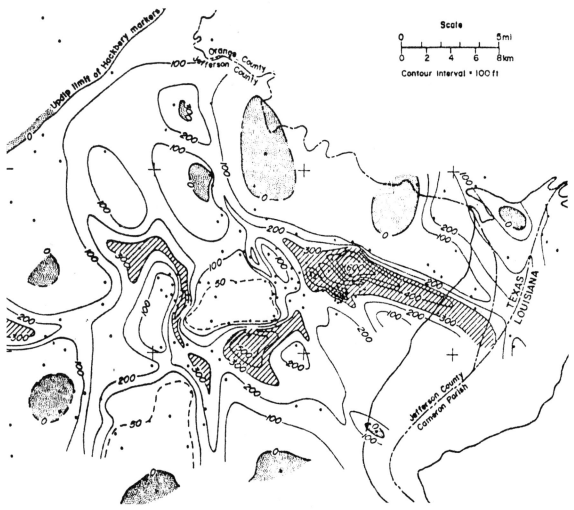

Figure 3. Net-sand isopach map of the western submarine channel complex, Hackberry slope system. From Ewing and Reed (in press).

ISOLATED SUBMARINE CANYONS--YOAKUM AND HARDIN CHANNELS OF THE WILCOX MARGIN

Large erosional gorges cut into paralic sediments are prominent features of many depositionally active, deltaic continental margins. These features have variously been interpreted as subaerial or submarine in origin. Evidence is generally considered as favoring submarine origin or at least significant submarine modification. Fill of the gorges consists of marine, commonly deep-water sediment.

Large well-known gorges of the Gulf Coast Eocene section occur within the lower and middle Wilcox Group (fig. 1). They include the Yoakum channel, first described by Hoyt (1959), Lavaca and subsidiary Smothers channels, described in recent papers by Chuber and Begeman (1982), and the Hardin channel. The Yoakum channel is 50 mi (80 km) long, extending from the Wilcox fault zone, which defines the position of the early Eocene paleocontinental margin, updip nearly to the present outcrop (fig. 1). It is more than 10 mi (16 km) wide and contains as much as 3,000 ft (900 m) of fill. Vormelker (1979) calculated that 74 mi^3 (312 km^3) of middle and lower Wilcox shelf, deltaic, and shore-zone sand and mudstone were excavated during canyon cutting. The gorge cross section is symmetrical (fig. 2). The gorge trends and deepens basinward across the stable Wilcox shelf platform, which was deposited updip of the buried Cretaceous reef trend. In contrast, the neighboring Lavaca channel is relatively broad and shallow. It lies within the progradational lower Wilcox depositional episode and extends only 12 mi (20 km) landward of the contemporary continental margin. Channel fill is about 1,000 ft (300 m) thick (Chuber and Begeman, 1982).

The stratigraphy of the Eocene gorges offers some insight into their possible origin. It is apparent in figure 1 that the channels lie along the margins of a major fluvial-dominated deltaic depositional system (described by Fisher and McGowen, 1967), in a transition zone between deltaic and shore-zone regimes. The Yoakum channel mudstone fill is shown to be correlative with a thin marine shale in the subsurface and is thought to be equivalent to the widespread Sabinetown transgressive shelf system. Gorge excavation entirely postdates lower Wilcox deposition, and nearly all of both the middle and lower Wilcox section is locally removed (fig. 1). The Lavaca channel appears to be correlative with a local shale bed that underlies a progradational deltaic sand section. Reflection seismic data can be reinterpreted to show an onlapped erosional seismic sequence boundary within the lower Wilcox.

Both Lavaca and Yoakum channel fills consist almost entirely of mudstone which grades up into progradational sequences capped by deltaic or shoreface sand bodies. Stratigraphic position, mud-dominated fill, and the scale of the features all indicate probable water depths ranging from neritic to bathyal.

Significantly, the deepest wells drilled to date have not penetrated sandy slope deposits. In these deeply incised transcontinental-margin sediment dispersal conduits, sand appears to have been bypassed to the base of the contemporary slope.

Key Reference: Chuber and Begeman (1982)

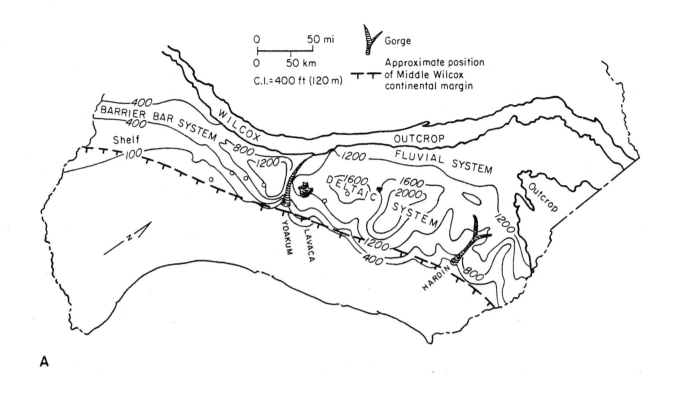

A

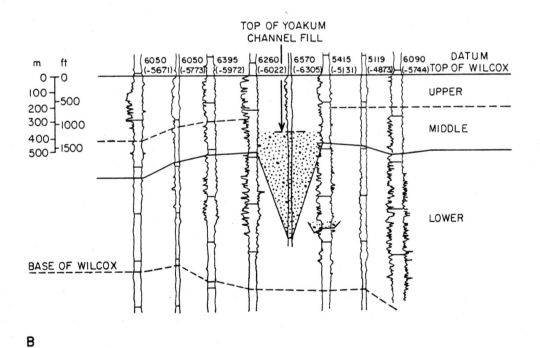

B

Figure 1. Sandstone isolith map and depositional systems of the lower Wilcox Group showing the depositional setting of the Yoakum, Lavaca, and Hardin gorges.

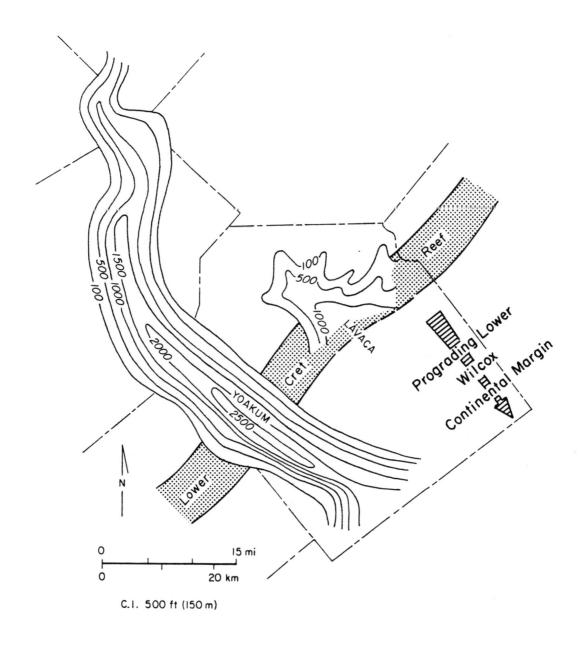

Figure 2. Isopach of the fill of the Yoakum and Lavaca gorges. Modified from Chuber and Begeman (1982).

SUMMARY--GULF COAST TERTIARY PALEOMARGINS

Criteria for Recognition

In the absence of large-scale depositional architecture that reflects the original shelf-slope break geometry, no single feature is diagnostic. However, the association of multiple criteria provides indirect evidence of the shelf-margin transition zone--a belt several miles wide in which "deep" and "shallow" water sands intermingle.

Tertiary Paleomargins -- Northwest Gulf Basin

Tertiary (and a few Cretaceous) paleomargins of the Northwest Shelf, as defined by the criteria summarized in table 1, are shown in figure 5. The basin rim is defined by the lower Cretaceous Edwards/Sligo reef trend. The lower coastal plain and continental shelf are products of large-scale depositional offlap of the continental margin, principally during Cenozoic time.

Key References: Winker (1982); Winker and Edwards (1983)

	CRITERION	PRINCIPLE	OBSERVED IN		COMMENTS
			LATE QUATERNARY	ANCIENT	
MOST PRECISE	topset-foreset geometry	stratification represents depositional relief of shelf and slope	X		usually obscured by contemporaneous structural growth or below resolution of conventional seismic data
	sedimentary structures	turbidites (in sandstones) and disrupted, chaotic bedding (in shales) characterize slope sediments	X	X	ancient slope sediments seldom cored in this type of basin; these sedimentary structures may also occur landward of the shelf edge.
	microfaunal assemblages Fig. 1	neritic-bathyal transition marks the shelf break	X	X	faunas can mix and interfinger; neritic faunas may be reworked downslope; depth assignments of assemblages not universally accepted by paleontologists.
	isopach maximum Fig. 2	maximum subsidence and sedimentation rates occur at shelf edge	X	X	usually insufficient downdip data to unequivocally demonstrate thinning basinward of shelf edge
	maximum rate of growth faulting Fig. 3	maximum extension rate and expansion ratios occur near shelf edge	X	X	usually insufficient deep data to observe pre-maximum fault growth
	stratigraphic top of geopressure Figs. 1 & 4	geopressure results from hydraulic isolation of sandstones against dewatering slope shales		X	relationship fairly circumstantial, but can be used in areas of very sparse data (mud weights and paleo tops)
LEAST PRECISE	"flexure"	sharp increase in regional dip and basinward-thickening rate mark relict shelf edge		X	shallow-water sediments commonly occur several miles downdip of contemporaneous flexure

Table 1. Criteria for recognition of ancient growth-faulted continental margins. The listing is in the approximate order of decreasing precision in locating the paleo-shelf break. From Winker (1982).

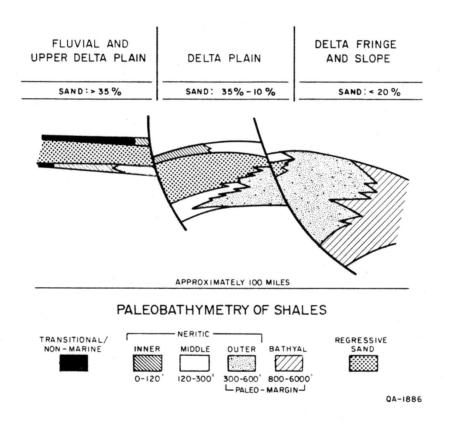

FLUVIAL AND UPPER DELTA PLAIN	DELTA PLAIN	DELTA FRINGE AND SLOPE
SAND : > 35 %	SAND : 35% - 10 %	SAND : < 20 %

APPROXIMATELY 100 MILES

PALEOBATHYMETRY OF SHALES

TRANSITIONAL/ NON - MARINE

NERITIC

INNER 0-120' MIDDLE 120-300' OUTER 300-600' BATHYAL 800-6000'

PALEO - MARGIN

REGRESSIVE SAND

QA-1886

Figure 1. Paleobathymetric framework of a Gulf Coast offlap depositional episode. Modified from Curtis and Picou (1980).

211

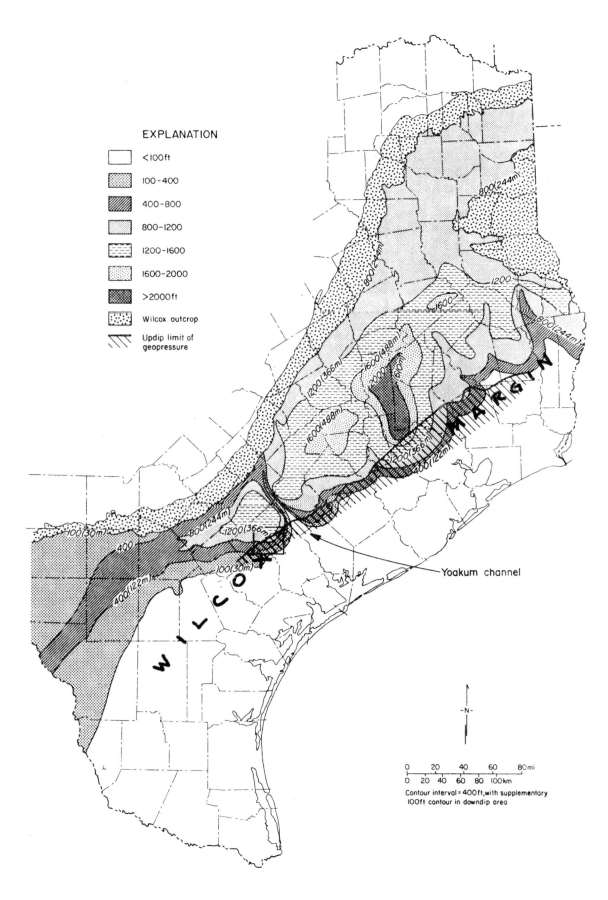

EXPLANATION

<100ft

100-400

400-800

800-1200

1200-1600

1600-2000

>2000ft

Wilcox outcrop

Updip limit of
geopressure

Yoakum channel

0   20   40   60   80mi
0 20 40 60 80 100km

Contour interval = 400ft, with supplementary
100ft contour in downdip area

Figure 2. Net sandstone distribution, updip extent of geopressure, and generalized position of paleomargin, lower Wilcox depositional episode. Modified from Bebout et al. (1982).

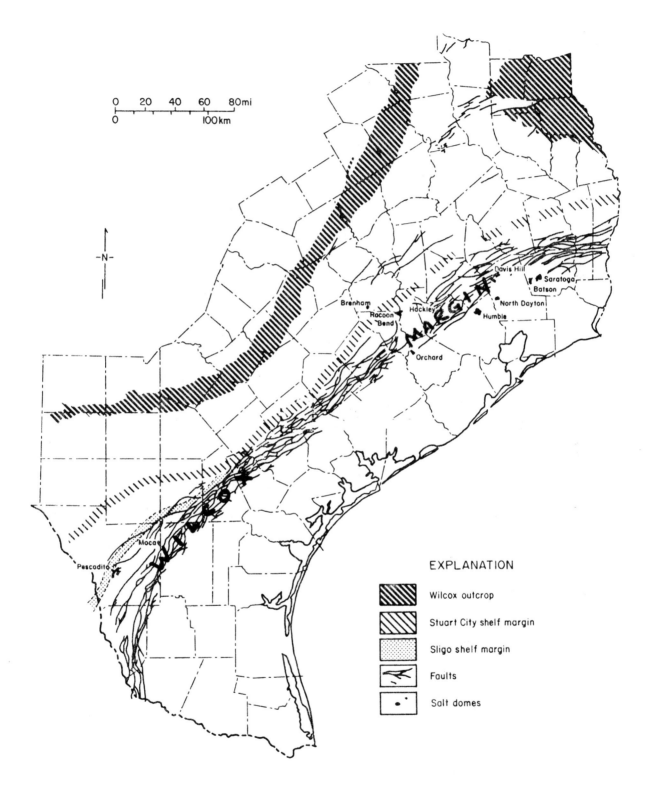

Figure 3. Structural framework of the Wilcox Group (undivided). The growth fault belt lies basinward of the underlying Cretaceous shelf edge. From Bebout et al. (1982).

213

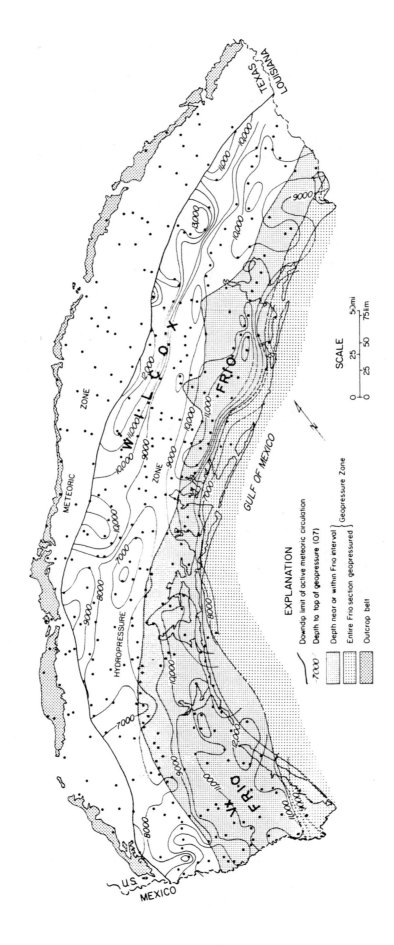

Figure 4. Hydrogeology of the Frio depositional episode. The top of geopressure is depressed beneath areas of sand-rich slope offlap (deltaic headlands) and pervades the Frio slope system downdip.

214

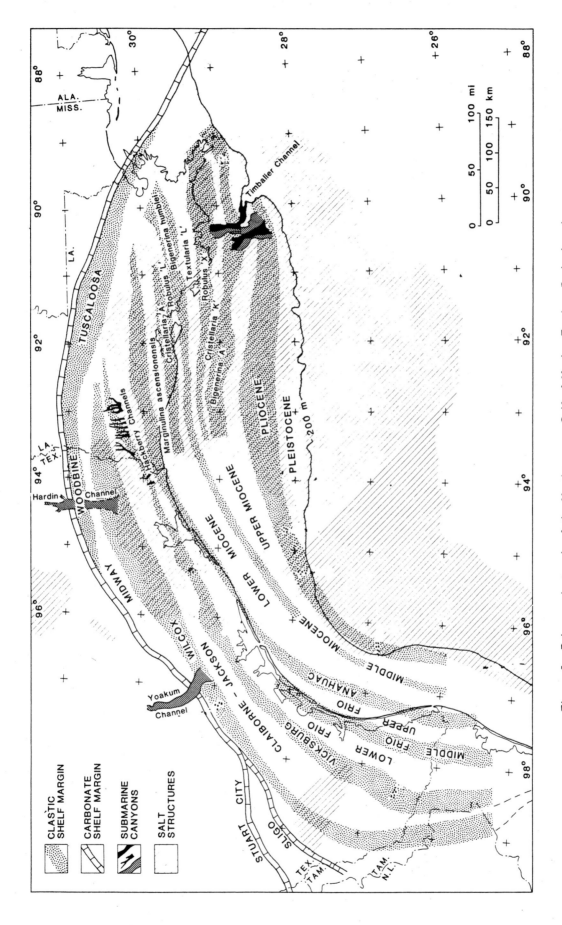

Figure 5. Paleomargin trends of the Northwestern Gulf of Mexico Tertiary Basin, based on application of the criteria listed in table 1. From Winker (1982).

PETROLEUM IN GULF CONTINENTAL MARGIN DEPOSITS

Tertiary paleomargin deposits commonly lie below burial depths of 10,000 ft. Although they occur within and above the oil window, they contain little oil. Typical production is overpressured gas and condensate.

Source Rocks

Continental slope deposits of the Quaternary and Tertiary are, by the conventional criteria for source rocks, of marginal quality and mixed oil and gas prone. They are however, the best sources available in the total facies tract (fig. 1).

Reservoirs

In addition to structural complexity, stratigraphic complexity is extreme in paleomargin sequences. Difficulty of correlation is characteristic of the upper slope. Potential reservoir facies include delta front (channel mouth bar, frontal splay, slump), distal shelf, and upper slope (onlap submarine canyon fill, megaslump, ponded turbidite) sand bodies (fig. 2). Onlap slope sands have provided the best reservoir quality. Distal delta/shelf and slope processes tend to mix sand, and silt/clay, adversely affecting reservoir quality. As gravitational processes play an increasing role in bed-load sediment transport, off-structure location of reservoir facies is accentuated. Critical question: To what degree are upper slope sands stratigraphically isolated from the feeder platform systems? In other words, is there really a new depositional system to prospect?

Generation and Migration

The isostatically balanced passive margin offlap superimposes several physical attributes of the basin fill within Tertiary paleomargin sequences (fig. 3). At typical thermal gradients (which increase in geopressured sediments), principal petroleum generation occurs in slope muds. Clay conversion and dehydration also occur mainly within margin sequences. Upward-directed hydrodynamic head, combined with the abundant structural conduits, leads to the attractive concept of large-scale vertical migration of generated hydrocarbons and chemically reactive water from the deep basin fill. Geochemistry of hydrocarbons within the Frio (fig. 4) supports the theory of large-scale injection of oil into the shallower deltaic and shore-zone reservoir sands. Gas and distillate within the geopressured paleomargin reservoirs may include thermally cracked liquids trapped in deeply buried reservoirs as well as a second generation of gas produced directly from slope source rocks. Again, thermal regimes and hydrocarbon geochemistry point to pervasive vertical migration.

Trapping

Paleomargin sediments are highly structured. However, complexity of deep structures, poor reservoir quality and continuity, and undependable (and evolving?) sealing capacity of fault planes place limits on the trap potential. An offlap slope is a mobilizer. Salt structures, which continue to grow after slope progradation has extended basinward, provide the major traps of the Gulf--salt domes and dome-related structures (fig. 5). Perhaps the greatest opportunity lies in more careful prospecting for flank pinch-outs.

Key References: Curtis and Picou (1980); Dow (1978); Galloway et al. (1982a); Roberts (1982)

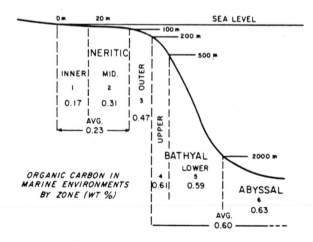

Figure 1. Organic carbon content of marine sediments by environmental zone (in weight percent). From Dow (1978).

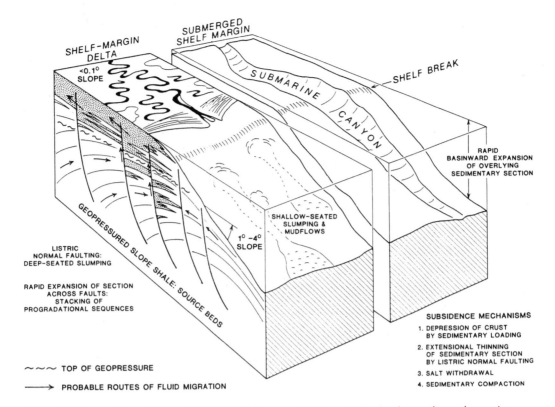

Figure 2. Reservoir settings and structural framework of unstable clastic continental margins. From Winker and Edwards (1983).

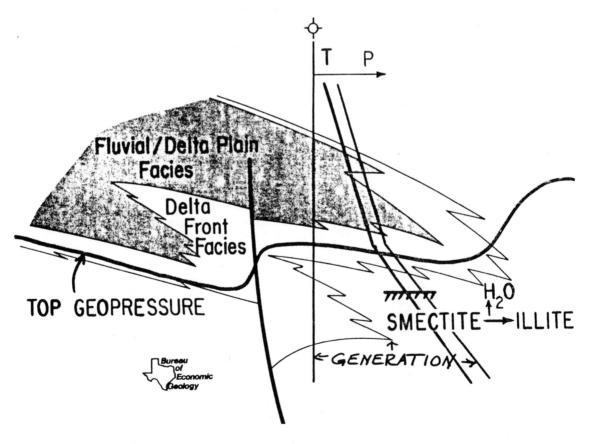

Figure 3. Convergence of temperature- and pressure-dependent geochemical and hydrologic processes within continental margin deposits.

219

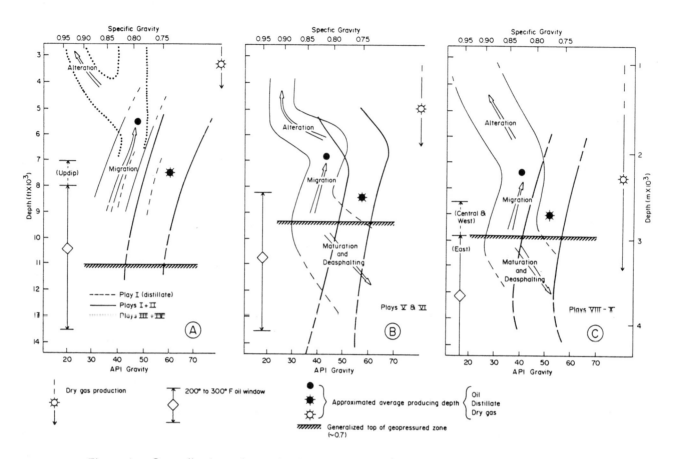

Figure 4. Generalized envelopes showing systematic variation in density of Frio oils and condensates with depth. Data are grouped into lower (A), middle (B), and upper (C) coastal plain plots. Well symbols indicate average depth of oil, condensate, and dry gas reservoirs in each area. Average vertical extent of the hydrocarbon liquid window and top of geopressure in each area are also indicated. From Galloway et al. (1982a).

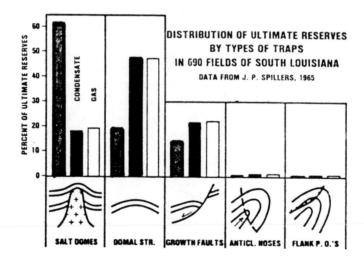

Figure 5. Distribution of ultimate reserves by trap type, south Louisiana. From Roberts (1983).

220

COMPARISON WITH OTHER DIVERGENT MARGINS

Divergent, or Atlantic-type, continental margin sags are a major class of petroliferous basins. Where such margins have been strongly influenced by large-scale deltaic progradation onto the slope, depositional and structural styles are very similar to those of the Tertiary Gulf Coast (fig. 1). Rapid depositional loading and consequent instability of the shelf edge and upper slope result in growth faulting and lateral extrusion of plastic shale and/or salt substrates.

Other than the Northwest Shelf of the Gulf of Mexico, the best known example of a delta-influenced margin is the Niger delta and its Tertiary basin. With Oligocene progradation to and beyond the edge of continental crust, thick sequences of delta-front and associated paralic Agdada Formation loaded the overpressured Akata slope mudstones (fig. 2). Growth faults created a variety of trapping structures, including simple and multiple growth faults with rollover anticlines, antithetic and counter-regional faults, and collapsed crests of shale ridges (Evamy et al., 1978). Along the east margin of the delta system, several large, mudstone-filled canyons and gulleys cutting through paralic and continental shales record episodes of submarine canyon excavation and filling (Burke, 1972).

Key Reference: Evamy et al. (1978)

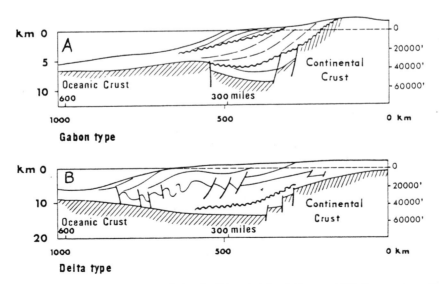

Figure 1. Schematic cross sections of simple (A) and delta-influenced (B) divergent continental margins. From Perrodon (1983).

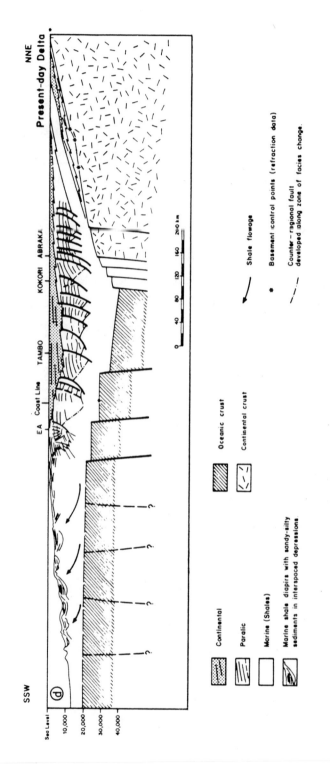

Figure 2. Simplified stratigraphic and structural cross section of the Cenozoic Niger delta system. From Evamy et al. (1978).

222

REFERENCES

Bebout, D. G., Weise, B. R., Gregory, A. R., and Edwards, M. B., 1982, Wilcox reservoirs in the deep subsurface along the Texas Gulf Coast, their potential for production of geopressured geothermal energy: The University of Texas at Austin, Bureau of Economic Geology Report of Investigations No. 117, 125 p.

Bishop, R. S., 1976, Shale diapirism and compaction of abnormally pressured shales in South Texas: Stanford University, Ph.D. dissertation, 168 p.

Bott, M. H. P., 1980, Mechanisms of subsidence at passive continental margins, in Bally, A. W., Bender, P. L., McGetchin, T. R., and Walcott, R. I., eds., Dynamics of plate interiors: Washington, D. C., American Geophysical Union, Geodynamics Series Volume 1, p. 27-35.

Brown, L. F., Jr., and Fisher, W. L., 1977, Seismic-stratigraphic interpretation of depositional systems: examples from Brazilian rift and pull-apart basins: AAPG Memoir 26, p. 213-248.

Bruce, C. H., 1973, Pressured shale and related sediment deformation: mechanism for development of regional contemporaneous faults: AAPG Bulletin, v. 57, no. 5, p. 878-886.

Buffler, R. T., Shaub, F. J., Watkins, J. S., and Worzel, J. L., 1979, Anatomy of the Mexican Ridges, southwestern Gulf of Mexico: AAPG Memoir 29, p. 319-327.

Carter, N. L., and Hansen, F. D., 1983, Creep of rocksalt: Tectonophysics, v. 92, p. 275-333.

Christensen, A. F., 1983, An example of a major syndepositional listric fault, in Bally, A. W., ed., Seismic expression of structural styles: AAPG Studies in Geology no. 15, v. 2, p. 2.3.1-36 to 2.3.1-40.

Coleman, J. M., and Prior, D. B., 1980, Deltaic sand bodies: AAPG Continuing Education Course Note Series no. 15, 171 p.

Crans, W., Mandl, G., and Haremboure, J., 1980, On the theory of growth faulting: a geomechanical delta model based on gravity sliding: Journal of Petroleum Geology, v. 2, no. 3, p. 265-307.

Curtis, D. M., and Picou, E. B., Jr., 1980, Gulf Coast Cenozoic: a model for the application of stratigraphic concepts to exploration of passive margins: Canadian Society of Petroleum Geologists, Memoir 6, p. 243-268.

Dailly, G. C., 1976, A possible mechanism relating progradation, growth faulting, clay diapirism and overthrusting in a regressive sequence of sediments: Canadian Petroleum Geology Bulletin, v. 24, no. 1, p. 92-116.

Dott, R. H., Jr., 1963, Dynamics of subaqueous gravity depositional processes: AAPG Bulletin, v. 47, no. 1, p. 104-128.

Dow, W. G., 1978, Petroleum source beds on continental slopes and rises: AAPG Bulletin, v. 62, p. 1584-1606.

Evamy, D. D., Haremboure, J., Kamerling, P., Knaap, W. A., Molloy, F. A., and Rowlands, P. H., 1978, Hydrocarbon habitat of Tertiary Niger delta: AAPG, v. 62, p. 1-39.

Ewing, T. E., 1983, Growth faults and salt tectonics in the Houston diapir province--relative timing and exploration significance: Gulf Coast Association of Geological Societies Transactions, v. 33, p. 83-90.

Ewing, T. E., and Reed, R. S., in press, Depositional systems and structural controls of Hackberry sandstone reservoirs, southeast Texas: The University of Texas at Austin, Bureau of Economic Geology Geological Circular.

Galloway, W. E., 1975, Process framework for describing the morphologic and stratigraphic evolution of deltaic depositional systems, in Broussard, M. L., ed., Deltas: Houston Geological Society, p. 87-98.

Galloway, W. E., and Hobday, D. K., 1983, Terrigenous clastic depositional systems: applications to petroleum, coal, and uranium exploration: New York, Springer-Verlag, 416 p.

Galloway, W. E., Hobday, D. K., and Magara, Kinji, 1982, Frio Formation of the Texas Gulf Coastal Plain--depositional systems, structural framework, and hydrocarbon distribution: AAPG Bulletin, v. 66, p. 649-688.

Galloway, W. E., Hobday, D. K., and Magara, Kinji, 1982, Frio Formation of the Texas Gulf Coastal Plain--depositional systems, structural framework, and hydrocarbon origin, migration, distribution and exploration potential: The University of Texas at Austin, Bureau of Economic Geology Report of Investigations No. 122, 78 p.

Gibbs, A. D., 1983, Balanced cross-section construction from seismic sections in areas of extensional tectonics: Journal of Structural Geology, v. 5, no. 2, p. 153-160.

Halbouty, M. T., 1979, Salt domes, Gulf region, United States and Mexico, Second edition: Houston, Gulf Publishing Company, 561 p.

Horsfield, W. T., 1980, Contemporaneous movement along crossing conjugate normal faults: Journal of Structural Geology, v. 2, no. 3, p. 305-310.

Kehle, R. O., in preparation, The origin of salt structures.

Lehner, P., 1969, Salt tectonics and Pleistocene stratigraphy on continental slope of northern Gulf of Mexico: AAPG Bulletin, v. 53, p. 2431-2479.

Lobeck, A. K., 1924, Block diagrams and other methods used in geology and geography: New York, John Wiley, 186 p.

Loocke, J. E., 1978, Growth history of the Hainesville Salt Dome, Wood County, Texas: The University of Texas at Austin, Master's thesis, 95 p.

Mandl, G., and Crans, W., 1981, Gravitational gliding in deltas, in McClay, K. R., and Price, N. J., eds., Thrust and nappe tectonics: Oxford, Geological Society of London, p. 41-53.

Martin, R. G., 1980, Distribution of salt structures, Gulf of Mexico: U.S. Geological Survey Map MF-1213.

Moore, G. T., Starke, G. W., Bonham, L. C., and Woodbury, H. O., 1978, Mississippi fan, Gulf of Mexico--physiography, stratigraphy, and sedimentational patterns: AAPG Studies in Geology no. 7, p. 155-191.

Morton, R. A., 1981, Formation of storm deposits by wind-forced currents in the Gulf of Mexico and North Sea: International Association of Sedimentologists, Special Publication 5, p. 385-396.

Paine, W. R., 1971, Petrology and sedimentation of the Hackberry sequence of southwest Louisiana: Gulf Coast Association of Geological Societies Transactions, v. 21, p. 37-55.

Pew, E., 1982, Seismic structural analysis of deformation in the southern Mexican Ridges: The University of Texas at Austin, Master's thesis, 102 p.

Ramberg, H., 1981, Gravity deformation and the Earth's crust in theory, experiments and geological application, second edition: London, Academic Press, 452 p.

Ritz, C. H., 1936, Geomorphology of Gulf Coast salt structures and its economic application: AAPG Bulletin, v. 20, no. 11, p. 1413-1438.

Roberts, W. H., 1982, Gulf Coast magic: Gulf Coast Association of Geological Societies Transactions, v. 32, p. 205-214.

Seni, S. J., and Jackson, M. P. A., 1983a, Evolution of salt structures, East Texas diapir province, part 1: sedimentary record of halokinesis: AAPG Bulletin, v. 67, no. 8, p. 1219-1244.

_____ 1983b, Evolution of salt structures, East Texas diapir province, part 2: patterns and rates of halokinesis: AAPG Bulletin, v. 67, no. 8, p. 1245-1274.

Sidner, B. R., Gartner, S., and Bryant, W. R., 1978, Late Pleistocene geologic history of Texas outer continental shelf and upper continental slope: AAPG Studies in Geology no. 7, p. 243-266.

Stude, G. R., 1978, Depositional environments of the Gulf of Mexico South Timbalier Block 54: salt dome and salt dome growth models: Gulf Coast Association of Geological Societies Transactions, v. 28, p. 627-646.

Talbot, C. J., 1978, Halokinesis and thermal convection: Nature, v. 273, p. 739-741.

Trusheim, F., 1960, Mechanism of salt migration in northern Germany: AAPG Bulletin, v. 44, no. 9, p. 1519-1540.

Walker, R. G., 1978, Deep-water sandstone facies and ancient submarine fans: models for exploration for stratigraphic traps: AAPG Bulletin, v. 62, p. 932-966.

Watkins, D. J., and Kraft, L. M., Jr., 1978, Stability of continental shelf and slope off Louisiana and Texas: geotechnical aspects: AAPG Studies in Geology no. 7, p. 267-286.

Weber, K. J., and Daukoru, E., 1975, Petroleum geology of the Niger Delta: Ninth World Petroleum Congress Proceedings, v. 2: London, Applied Science Publishers, p. 209-221.

Wernicke, B., and Burchfiel, B. C., 1982, Modes of extensional tectonics: Journal of Structural Geology, v. 4, no. 2, p. 105-115.

Winker, C. D., 1979, Late Pleistocene fluvial-deltaic deposition, Texas Coastal Plain and shelf: The University of Texas at Austin, Master's thesis, 187 p.

_____ 1982, Cenozoic shelf margins, northwestern Gulf of Mexico Basin: Gulf Coast Association of Geological Societies Transactions, v. 32, p. 427-448.

Winker, C. D., and Edwards, M. B., 1983, Unstable progradational clastic shelf margins: Society of Economic Paleontologists and Mineralogists, Special Publication No. 33, p. 139-157.

Winker, C. D., Morton, R. A., Ewing, T. E., and Garcia, D. D., 1983, Depositional setting, structural style, and sandstone distribution in three geopressured geothermal areas, Texas Gulf Coast: The University of Texas at Austin, Bureau of Economic Geology Report of Investigations No. 134, 60 p.

Withjack, M. O., and Scheiner, C., 1982, Fault patterns associated with domes--an experimental and analytical study: AAPG Bulletin, v. 66, no. 3, p. 302-316.

Woodbury, H. O., Murray, I. B., Jr., and Osborne, R. E., 1980, Diapirs and their relation to hydrocarbon accumulation, in Miall, A. D., ed., Facts and principles of world petroleum occurrence: Calgary, Canadian Society of Petroleum Geologists, p. 119-142.

Woodbury, H. O., Spotts, J. H., and Akers, W. H., 1978, Gulf of Mexico continental-slope sediments and sedimentation, in Bouma, A. H., Moore, G. T., and Coleman, J. M., eds., Framework, facies, and oil-trapping characteristics of the upper continental margin: AAPG Studies in Geology No. 7, p. 117-153.